By eating the fourteen **SuperFoods** highlighted in Dr. Steven Pratt's groundbreaking bestseller, you can actually halt the incremental deteriorations that lead to common ailments and diseases:

- *Beans*—help reduce obesity
- *Blueberries*—lower the risk for cardiovascular disease
- *Broccoli*—lowers the incidence of cataracts and fights birth defects
- *Oats*—reduce the risk of type II diabetes
- *Oranges*—help prevent strokes
- *Pumpkin*—lowers the risk of various cancers
- *Wild salmon*—lessens the risk of heart disease
- *Soy*—lowers cholesterol
- *Spinach*—decreases the chance of cardiovascular disease and age-related macular degeneration
- *Tea*—helps to prevent osteoporosis
- *Tomatoes*—raise the skin's sun protection factor
- *Turkey*—helps build a strong immune system
- *Walnuts*—reduce the risk of developing coronary heart disease, diabetes, and cancer
- *Yogurt*—promotes strong bones and a healthy heart

SUPERFOODS Rx

You'll feel super!

Books by
Steven G. Pratt and Kathy Matthews

SUPERFOODS HEALTHSTYLE
SUPERFOODS RX

SuperFoods Rx

Fourteen Foods That Will Change Your Life

STEVEN G. PRATT, M.D., AND KATHY MATTHEWS

**Recipes by Michel Stroot of the Golden Door
and the staff at Rancho La Puerta**

HARPER

An Imprint of HarperCollinsPublishers

HARPER

An Imprint of HarperCollins*Publishers*
10 East 53rd Street
New York, New York 10022-5299

Copyright © 2004 by Steven G. Pratt and Kathy Matthews, Inc.
ISBN: 978-0-06-117228-1
ISBN-10: 0-06-117228-6

First Harper paperback printing: January 2007
First Harper Paperbacks printing: November 2005
First William Morrow hardcover printing: January 2004

HarperCollins® and Harper® are trademarks of HarperCollins Publishers.

Printed in the United States of America.

Visit Harper paperbacks on the World Wide Web at www.harpercollins.com

10 9 8 7 6 5 4

In memory of Alex Szekely

Contents

....

PART III ■ THE SUPERFOODS RX MENUS AND NUTRITIONAL INFORMATION

Acknowledgments

I'd like to thank my entire family—my wife, Patty; my kids, Mike, Ty, Torey, and Brian; and Mike's wife, Diane—for their time, effort, and patience in the course of this project. It has truly been a family effort. I would like to thank my partners, David Stern and Ray Sphire, who believed in this project from the start and have made significant contributions to the development and publication of this book. I would also like to thank my patient and friend Nancy Stanley for her healthy, fun recipe, and Michelle McHose from my office, who did lots of research. I owe a thank you to everyone in my office, especially Carol Henry and Maurya Hernandez, for helping me to keep my clinical practice running smoothly while I spent time out of the office working on the book.

Many thanks also to my good friend and colleague Dr. Hugh Greenway for his vision of this book and for his dedication and involvement in our original idea—the SPF Diet—which we will continue to work on, expanding our knowledge and research efforts in our goal to use nutrition and lifestyle changes to prevent disease.

We cannot thank enough the group of committed individuals at the Golden Door and Rancho La Puerta who took on a project that their leader Alex Szekely believed would have an impact on the health and lives of others. Deborah Szekely, founder of the Rancho La Puerta and the Golden Door, has been a source of great encouragement and inspiration.

I must particularly thank Mary-Elizabeth Gifford, director of communications for Rancho La Puerta and the Golden Door, for her unwavering support and for her generosity of time and wisdom toward all of us on the SuperFoods Rx team, and Michel Stroot, executive chef at the Golden Door, who has made a giant contribution to our book and whose vision and direction has resulted in recipes that clearly demonstrate you can find wonderful ways to prepare the healthiest ingredients in your own home. Everyone should try to find the time to experience Chef Michel's food in his own environment at the Golden Door or Rancho La Puerta. Also working with Chef Michel on this project was Dean Rucker, sous chef at the Golden Door. Dr. Wendy Bazilian, Dr.P.H., M.A., R.D., nutrition consultant and registered dietitian to the Golden Door, who is one of the busiest people we know and who nevertheless made time to develop and coordinate recipes that have integrated all the SuperFoods Rx ingredients and nutritional guidelines. Yvonne Nienstadt, director of nutrition at Rancho La Puerta, and Chef Gonzalo Mendoza contributed creative recipes and insights so that SuperFoods Rx will be rewarding to its readers. Thanks to Mary Goodbody for expert recipe testing and to Lori Winterstein, registered dietitian, for her hard work analyzing the nutrient content of each of the ten-day menu plans.

Thank you to Dr. Gary Beecher for his expertise and advice on the nutritional content of certain foods, and Dr. Joe Vinson for his excellent job analyzing the polyphenol content of various juices and jams. I also appreciate the help and data Dr. Eric Van Kuijk, his brother Bas, and Jair

Haanstra provided on the nutritional content of sweet orange bell peppers (a definite whole food tasty "all star").

I am fortunate to have as my good friend Dr. Stewart Richer, who has shared his unparalled clinical nutritional knowledge with me over the last three years. I would also like to acknowledge the extensive carotenoid data and research that Doctors Norman Krinsky, Max Snodderly, Billy Wooten, and Billy Hammond have been so kind to share with me.

Thanks to the world's greatest agent and now good friend, Al Lowman. Thanks to Kathy Matthews and Harriet Bell for their dedication, hard work, friendship, and expertise. They surely are, along with Al, the greatest team an author could dream of.

A very big thank you to the entire HarperCollins crew: Lisa Gallagher, our marketing guru; Heather Gould, publicist extraordinaire; Roberto de Vicq de Cumptich, our extraordinary jacket designer; Michael Morrison, our publisher and champion; and Sonia Greenbaum, our intrepid copyeditor.

—STEVEN G. PRATT, M.D.

My family has exhibited endless patience and good humor in the course of the preparation of this manuscript, eating frozen entrees while I wrote about the remarkable benefits of whole, fresh foods. To them—my husband, Fred; my sons, Greg and Ted—a huge thanks and a promise of many future SuperFood meals!

Steve Pratt has been the most extraordinary colleague—dedicated, cheerful, relentlessly hardworking, and stunningly well-informed and articulate in all matters relating to nutrition. Many thanks to him and his delightful wife, Patty, for making this project fun. David Stern, Ray Sphire, and Hugh Greenway have been constant and enthusiastic friends to this book and I am very grateful to them for their support.

Al Lowman, agent and friend for a very long time, has once again proved his mettle as ally extraordinaire. He is one in a million.

Harriet Bell, one of the best editors ever, and the entire team at William Morrow have been truly wonderful. Their support, energy, meticulous attention, and enthusiasm have been an absolute delight and I am grateful to them all.

The people at Rancho La Puerta and the Golden Door have been the secret ingredient in this book, making it a very special experience. I now understand why people return to both spas time and time again. I am grateful to them for their hospitality to me, their spirit of wellness, and their skill at turning a batch of greens into a memorable and delicious meal. I would especially like to thank Mary-Elizabeth Gifford for her early guidance and enthusiasm, Chef Michel Stroot and nutritionists Yvonne Nienstadt and Wendy Bazilian, and Deborah Szekely, founder of Rancho La Puerta and the Golden Door, who is a pioneer and an inspiration in finding a healing spirit in everyday life.

—KATHY MATTHEWS

Foreword

With the advent of fast-food chains and TV dinners, the whole idea of fresh vegetables and healthy foods was essentially forgotten by most of America. But little by little we are seeing a resurgence of interest in and commitment to healthy foods with greenmarkets springing up and organic foods being sold in supermarkets. *SuperFoods Rx* is the next major step forward in the food-health connection, and I am certain that people are ready to appreciate that what we eat can act as good or bad "medicine."

While most of us are well aware that food choices are important to health, very few people realize that foods can actually *promote* health. This is good news to us all because it means that instead of worrying about what foods we should avoid, it introduces an era when we can, at last, embrace foods—particularly SuperFoods—as allies in our quest to live long, healthy lives.

We are constantly increasing our awareness and knowledge that "we are what we eat." The *Wall Street Journal* recently featured articles including "Yes, There Are Some Healthy Snacks That Your Kids Will Actually Eat," "Toward Smarter Snacks," and "Big-Brand Logos Pop Up in Organic Aisle." The very next day in the same publication was a half-page color spread of fruits and vegetables by a major food company welcoming the message "Diets rich in fruits and vegetables may reduce the risk of some types of cancer and other chronic diseases (July 10, 2003, U.S. National Cancer Institute, as endorsed by the U.S. Food and Drug Administration)."

SuperFoods Rx: Fourteen Foods That Will Change Your Life is an exciting concept based strongly in academic science and peer-reviewed research clinical trials conducted by some of the world's leading clinicians and research scholars. My close friend Steve Pratt, M.D., has devoted many hours to nutrition and its science and implementation, and this book is the exciting culmination of those efforts.

Steve and I have been practicing physicians, colleagues, and close friends in southern California for more than twenty years. In addition to being practicing specialists, we also are interested in the entire patient. My original background was in family medicine where I first became board certified in family practice. Steve is also a physician interested in the entire well-being of his patients. Thus, the inquiry and journey from treatment to prevention to concern for the whole patient has always been an important part of our medical practices. As the former CEO of the world-renowned Scripps Clinic, I can state that physicians in all of our specialties are gaining an increasing awareness of the role that nutrition and foods play in our patients and their disease processes. I think the important underlying concept of *SuperFoods Rx*—that certain foods can promote and enhance health—will be embraced by all health care practitioners as well as the general public. It's a book whose time has come.

The SuperFoods Rx Lifestyle Pyramid is of particular interest to those looking for a balanced, healthy lifestyle. It includes more than just food. Steve correctly includes exercise, stress management, faith, friendship, laughter, daydreaming, sleep, and other elements necessary for a long-standing, healthy life.

I am pleased to be continuing to work with Dr. Pratt in all aspects of medicine, especially our research efforts. Together we are looking at what foods might function as natural sunscreens in the plant world and how such information can help and protect us in the future. Both the eye and the skin provide a window into the inner aspects of our bodies, so our research is important for a multiplicity of reasons.

Change can be both challenging and difficult for all of us. To think that overnight you and I would fully incorporate the fourteen SuperFoods into our daily routine may be unrealistic. Here's how I plan to do it: I plan to add one SuperFood a week to my life in 2004. Call me in January 2005 and I bet I'll both be and feel healthier than ever!

Steve remains at the forefront of helping each of us live better, longer, and more productive lives with his research, leadership, and forward thinking in nutritional medicine. I hope that *SuperFoods Rx* will help you find a better, healthier future.

—HUBERT (HUGH) T. GREENWAY, M.D.,
CEO emeritus, Scripps Clinic,
chairman, Mohs Dermatologic Surgery

Introduction

WELCOME TO SUPERFOODS RX

Each time you sit down to a meal, you're making life-and-death decisions. Does that sound scary? At first blush it truly is. But here's how I see it: you have an exciting opportunity—one that wasn't available even a few years ago—to make choices that will change the course of your lifespan and your health span. You are making decisions right now, at your very next meal, that will affect how you spend the rest of your life, whether you're 22 years old or 62.

See this book as a fork in the road:

One way leads to a handicapped space in front of the mega-drugstore. You're 68 years old and you're struggling to navigate to the back of the store where the pharmacist has your meds waiting. He greets you cheerfully; he knows you well. You're taking nine prescription medications and on a good day, you can walk around the block with your grandson. On a bad day, you log a lot of TV time. You fill your basket with your prescriptions and a few over-the-counter medicines. You shake your head sadly when the elderly man behind you in line says, "Old age isn't for the faint of heart, is it?"

There's another choice. You pull up to the farmers' market. You're 68 years old and you've just finished a tennis game with some pals or an hour of satisfying work in your garden. You grab a basket and fill it with delicious fruits and vegetables. You're thrilled to see that blueberries have come into season and you stock up on spinach and orange bell peppers. There are no ripe tomatoes yet but the broccoli looks perfect. You've got friends coming over for dinner so you'll stop by the fish market for some wild Alaskan salmon. Maybe you'll pick up some almonds; you've been meaning to try that new dessert recipe. You duck out of line for a moment to grab a small red cabbage and smile when the cashier says, "I don't know where you get your energy!"

It's a simple choice, really: the right foods or prescription drugs.

Of course no one can guarantee that one will ensure you'll avoid the other. But there is enough evidence—some published, some just being reported at medical conferences—that the power of certain foods can make a significant difference in your risk of developing a host of diseases. This is extremely exciting because it puts the tools in your hand, and on your plate, to change your future.

THE FABULOUS FOURTEEN

SuperFoods Rx is based on a very simple concept: some foods are better than others for your health. We could all guess that an apple is better for you than a potato chip. But what about choosing between a couple of pretzels and a few walnuts? Did you know that eating a handful of nuts a few times a week can reduce your risk of getting a heart attack by at least 15 percent and perhaps as much as 51 percent? Even if you smoke, are overweight, and never exercise. That's how powerful certain foods are.

SuperFoods Rx presents the fourteen known nutritional powerhouse foods that can help you extend your health span—the extent of time you have to be healthy, vigorous, and vital—as well as, perhaps, your lifespan. These are the foods that have been proven to help prevent and, in some cases, reverse the well-known scourges of aging, including cardiovascular disease, type II diabetes, hypertension, certain cancers, and even dementia.

Many SuperFoods may already be part of your diet. Most people routinely enjoy **broccoli** and **oranges** and even **spinach**, for example. **Blueberries** could possibly be everyone's favorite SuperFood, but most people only eat them as a treat, when they're in season. When you learn about the awesome power of this berry, and some other berries, to promote health I'm sure that, like me, you'll try to eat berries every day. Frozen berries in a shake are delicious. Some SuperFoods, like **pumpkin** and **turkey**, make occasional appearances in the average diet. (You'll soon see why they should be eaten far more frequently.) Other SuperFoods may be brand-new to you, like **soy** or perhaps **yogurt**, and I'll show you how to enjoy them even if you think you don't like them.

Walnuts are the surprising SuperFood. Surprising because most people think nuts belong in the "avoid" category because they're fatty. But the power of nuts is almost stunning. I try never to go a day without eating some nuts and seeds. Some SuperFoods are less of a surprise. Perhaps you've read about **wild salmon** recently. Wild salmon has so many health benefits that it seems just foolish not to eat it regularly.

Some SuperFoods are so easy to incorporate into your diet that you'll be able to begin your overall improvement in the time it takes to boil some water for a cup of **tea** to sip while you finish reading this book. Others will take a little more planning. Some SuperFoods, such as pumpkin (or one of its sidekicks, like orange bell pepper), should be eaten a few times a week; others, like yogurt, should be eaten more frequently, even if in small amounts. A couple of Super-Foods cover a category—for example, **beans**, which you may rarely think about—but when you realize the power of a humble pea or garbanzo to improve your health, you'll make a point of eating them frequently. And when you discover the power of **tomatoes**, even in the form of ketchup or sauce, you'll find yourself choosing pizza over other fast foods.

You may be surprised to learn from SuperFood shopping tips that a dramatic improvement in your nutrition can be a simple matter of learning to read a label. For example, most of us eat bread every day and we serve it to our families. Many of us think we're eating **whole grain** bread and would be surprised to learn that we're not. Did you know that by checking the nutrition label on a loaf of bread and making sure it has at least 3 grams of fiber, you could turn an ordinary sandwich into a SuperFoods Rx sandwich?

Most of us tend to get set in our ways: we eat the same things on a fairly regular basis. To break out of your particular rotation, you need convincing that it's worth it and also that it's easy to do. I know that when you read about the individual SuperFoods you'll be convinced that it's worth eating them. To make it easy, *SuperFoods Rx* gives you a hand in the kitchen. Each chapter has lots of tips and suggestions that will help you choose and prepare each Super-Food. I've also included my favorite, simple, everyday family recipes—most developed by my wife, Patty—that will get you right on the SuperFood way to health. In addition, one of the best spa chefs in the country, Michel Stroot, chef at the Golden Door in California, was given free rein to develop some great-tasting recipes using SuperFoods.

Some of these recipes are quick and easy; others are more suitable for a splurge. They're all absolutely delicious, and perhaps the most healthy recipes you'll find anywhere.

To make it all super easy, I've created shopping lists that will direct you right to the best products our markets have to offer. There arc excellent, healthy, prepared foods out there if you just know what to look for. Of course, most of us don't have time to read and compare countless labels in the supermarket. You don't have to: I've done it for you. Check the SuperFoods Rx Shopping Lists (page 324) and you'll be delighted to find another tool to help you improve your everyday diet. My patients love these lists and I'm sure you will too.

FEELING SUPER

Remember that *SuperFoods Rx* is not just about avoiding disease. The fact is, few people are willing to make changes in how they live every day with the hope that thcy won't develop a disease in a few decades. But it's not just the last quarter or third of your life that's affected by making the wrong food choices; it's every day. Once you're beyond those enviable, vigorous teen years, you hear your "health clock" beginning to tick. Small aches show up. You feel tired in the late afternoon. You don't have much enthusiasm for a bike ride or any vigorous activity, for that matter. Your skin loses some of its glow. In many cases these minor symptoms are the easily dismissed, early-warning signs of what can develop into future chronic ailments.

I often tell my patients, "I wish you could feel like I feel." I'm not boasting. I'm encouraging them to alter their eating patterns because I know that once they begin to feel really good, that will be all the encouragement they'll need to make their changes permanent.

Many people find it hard to believe that they could be developing disease if they feel okay and have reasonably good health habits. Sadly, if this were true, we wouldn't have the host of chronic and disabling diseases that we're observing today.

Take this quick test: get a mirror, preferably a magnifying mirror, and examine your eyes. Look at the white part, particularly the area of the eye just to the left or the right of the iris—the colored part. Do you see some yellow discoloration? Sometimes it can look like yellowish ropey globs. These are called "pinguecula." They're a kind of callus that the thin mucous membrane that covers the eye develops as a result of exposure to pollutants and, in particular, ultraviolet light. Perhaps you also notice a yellow ring around the peripheral part of your cornea (the front, clear part of the eye). This is a warning sign that your cholesterol may be high and you should have a blood test to check your cholesterol levels. If you have one or both of these signs, your body is telling you that your diet and your environment may be taking a toll on your immune system and your eventual health.

Foods—the right foods—can actually change the course of your biochemistry. They can help to stop damage at the cellular levels that can develop into disease. The goal of *SuperFoods Rx* is to help you stop the incremental changes in your body that can lead to disease and/or dysfunction. The delightful side effect to this effort is that you feel better, have more energy, look better, and can embrace all that life has to offer you with more optimism.

THE BEGINNINGS OF SUPERFOODS

My mom introduced me to the concept of optimum nutrition. I had no choice: I had to eat the crusts on the sandwich (we now know they really are the healthiest parts). I ate salads every day. There was always a jar of wheat germ in our fridge and I was encouraged to eat the white part of the orange peel. My mom was an early devotee of nutritionists like Adele Davis. While I might have resisted some of her efforts as a kid, when I became a serious athlete and a doctor, I saw that she was right.

As an athlete I was aware that nutrition affected performance. It was clear that what I ate affected my ability to compete. Gradually, I became interested in how foods affected performance on a biochemical level: what was there about certain foods that made them especially beneficial?

As an ophthalmologist and a plastic and reconstructive surgeon, I found myself working in an area of medicine that gave access to the body's first warning signals of age and disease: the eyes and the skin. You can't see your arteries narrowing. But you can experience your eyes losing clarity and developing pinguecula, and you can see your skin developing discolorations and losing elasticity.

Skin health, ocular health, and overall health are inextricably linked. For example, if you're diagnosed with cataracts or macular degeneration, you have a significantly increased risk of developing cardiovascular disease. It's the same story with skin cancer. If you develop skin cancer before the age of 60, you may have a 20 to 30 percent increased mortality rate from multiple systemic cancers, including colon cancer, breast cancer, prostate cancer, and leukemia.

How I've Incorporated SuperFoods into My Life

I've always followed a version of the *SuperFoods Rx* diet, which gets updated regularly as I learn new things and new research reports are published. People who are involved in nutrition research usually adopt changes in their own lives based on what they learn. For example, the people who did the analysis for me on various fruit juices told me that after learning how concentrated the antioxidants are in certain juices, they make a point of drinking them frequently. I thought you'd find it interesting to get a quick snapshot of some of the changes I've made in my own diet as a result of working on *SuperFoods Rx*.

- In my fridge there are separate plastic containers for dry-roasted sunflower seeds, almonds, walnuts, dry-roasted peanuts, pistachios, and pumpkin seeds. I have a handful of at least two different nuts and/or seeds every day.

- I drink one Odwalla C Monster juice almost every day. I sip it slowly over hours instead of drinking it all at once to keep my vitamin C blood levels high.

- I put lots of either Knott's Boysenberry Preserves or Trader Joe's Organic Blueberry Fruit Spread on my toast. My kids used to tease me about the quantity of Knott's on my toast, but I now have scientific proof that this was always a good idea.

- I drink a lot of green and black tea. My drink of choice when dining out is iced tea, especially if the restaurant brews the tea instead of using instant. I always squeeze some fresh lemon into the tea.

- I usually have 5 ounces of Trader Joe's 100% Unfiltered Concord Grape Juice or pomegranate juice with sparkling water with my dinner or lunch at home.

- I eat a cup of berries almost every day.

- I put some ground flaxseed on my cereal.

- The SuperFoods Rx Salad (page 273) is a daily must.

- I always check food labels for sodium content.

My practice at the Scripps network of clinics and hospitals in San Diego, California, provides a wealth of clinical material. Most research centers struggle to get patients. Many researchers aren't really clinicians. They don't work with patients and they don't get to see real results in real people. I'm a practicing clinician who also does research. I can draw on my clinical practice to study anything. The Scripps health care system is world famous and we draw an excellent and cooperative patient base. In addition, I'm surrounded at Scripps by researchers who are on the cutting edge of their fields, researchers who are only steps away when I have a question or need clarification on an arcane point of biochemistry or medicine. The combination of top-notch experts and world-class library and research facilities at UCSD (University of California at San Diego), where I'm a member of the clinical teaching faculty, has been a tremendous advantage to me in my work. It's given me access to the cutting-edge advances in nutritional therapy.

It was serendipitous that these interests—nutrition, ocular and skin health, and preventive medicine—meshed. It quickly became clear to me from clinical research, as well as from my own work with patients, that the connections between food and specific nutrients and health were inevitable.

I also realized that the public, despite a hunger for such information, was finding it increasingly difficult to find simple, safe, dietary recommendations. The public thirst for knowledge has, ironically, been part of the problem. People are eager to hear about the latest advance. Nutrition updates hit the front pages. Unfortunately, oftentimes headlines don't translate into sensible, practical recommendations. People become discouraged by conflicting information. They begin to think that there's no point in trying to improve their diet; what they're told to eat today turns out to be a problem tomorrow.

I have taken the best of what's known about the best foods available in order to show you how to develop an "optimum" diet. *SuperFoods Rx* will help you fine-tune your diet to get the most out of the good foods you already eat, while introducing other, powerful, health-promoting foods to amplify the overall beneficial effects. *SuperFoods Rx* will be your road map to a better, healthier future.

PART I

SuperFoods Rx: The Basics

How Your Diet Is Killing You

The foods you eat every day, from the fast food you mindlessly consume to the best meals you savor in a top restaurant, are doing much more than making you fat or thin. Their effects on your body are making the difference between the development of chronic disease and a vigorous extended life. They can prevent or greatly reduce your risk of vision problems, stroke, heart disease, diabetes, and a host of killers. These are not just vague promises; they are facts that are now supported by an impressive and irrefutable body of research.

Most respectable scientists in the world today agree that at least 30 percent of all cancers are directly related to nutrition. Some would argue that the figure is as high as 70 percent.

For example, we know that the people who eat the most fruits and vegetables are half as likely to develop cancer as those who eat the least amount of these foods.

It's not just cancer that's nutrition related: about half of all cardiovascular disease and a significant percentage of hypertension cases can be traced to diet as well. In the Nurses' Health Study (an ongoing study of over 120,000 female nurses, begun in Framingham, Massachusetts, in

1976), the nonsmoking women with a median daily intake of 2.7 servings of whole grains were half as likely to suffer a stroke as other women in the study. Given this, it's particularly alarming to learn that fewer than 8 percent of Americans eat this much whole grains.

Indeed, most of us are eating ourselves to death: only about 10 percent of Americans eat the foods that would enable them to be free of chronic disease and premature death.

Our Western diets are literally killing us. While man evolved on a plant-based diet more than fifty thousand years ago, our modern diet—the one our parents ate and the one we're eating—developed only during the past fifty to eighty years. It is not serving us well. We humans are genetically "wired" for starvation, not an overabundance of food. Our genes are set for hunter-gatherer mode, and a diet rich in fruits, vegetables, whole grains, nuts and seeds, and lean, wild game, not for the majority of foods and beverages found in today's supermarkets.

• •

It has been estimated that 300,000 to 800,000 preventable deaths per year in the United States are nutrition related. These include deaths from atherosclerotic disease, diabetes, and certain cancers.

• •

Here are eleven disastrous developments in nutrition that are ruining your health and the health of most everyone in modern industrialized societies:

1. Increased portion sizes.
2. Decreased energy expenditure; people just don't exercise enough.

3. Unhealthy balance of fats in the diet: an increase in saturated fat, omega-6 fatty acids, and trans-fatty acids, along with a huge decrease in omega-3 fatty acids.
4. An increase in consumption of processed cereal grains.
5. An overall decrease in fruit and vegetable intake from historical standards.
6. A decrease in lean meat and fish intake.
7. A decrease in antioxidant intake and calcium intake (especially from whole foods).
8. The unhealthy ratio of omega-6 to omega-3 fats, which is associated with a long list of chronic diseases.
9. A marked increase in refined sugar as an overall percentage of caloric intake.
10. A decrease in whole food consumption, which has led to a marked decrease in phytonutrient intake.
11. A decrease in the variety of foods eaten.

Few people, including health professionals, are aware of the significant recent decline in our overall health status. More than 125 million Americans have at least one chronic condition like diabetes, cancer, heart disease, or glaucoma. The Centers for Disease Control estimates that one-third of Americans who were born in 2000 will develop diabetes in their lifetime. Sixty million Americans have more than one condition. It's getting worse every day. In 1996, estimates were made projecting the rate of chronic disease in the future.

Four years later, in 2000, the number of people with chronic ailments was twenty million *higher* than had been anticipated. By the year 2020, a projected one-quarter of the American population will be living with multiple chronic conditions, and estimated costs for managing these conditions will reach $1.07 trillion.

The most shocking nugget of information in this dismal overview of American health is that the age of the "chronically ill" is declining. About half of chronically ill Americans are under age 45 and, stunningly, 15 percent of that number are children who are suffering from diabetes, asthma, developmental disabilities, cancer, and other disorders.

As a doctor, I see the imperfections of the system every day. The general unspoken assumption among many people is that you can eat whatever you feel like eating and count on a pill or a surgery to take care of the fallout down the line. For many of us, the only diet-related concern, if we have one, is weight control.

••

Recognizing the crisis in pediatric health care, for the first time in the spring of 2003, the American Heart Association has offered guidelines for screening kids.
They include:
- Check a child's blood pressure at every visit after age 3.
- Talk to kids about not smoking as early as age 9.
- Test cholesterol levels and blood fats in kids who are overweight or at risk.
- Review family history for signs of early heart disease.

••

••

Two of three U.S. adults are either overweight or obese compared with fewer than one in four in the early 1960s. Obesity accounts for more than 280,000 deaths annually in the U.S.

••

What's the answer? Clearly, we need to do better if we want to live longer and avoid chronic disease. In simplest terms, we need to work with a system—our bodies—that's geared to thrive in times of starvation and high-energy expenditure and adjust for greatly reduced activity levels in a world where food is overabundant. In other words, *we need to get as much nutrition as possible from fewer calories*. This is possible only if we select the most nutrient-dense, low-calorie foods and make these foods the backbone of our daily diet. *SuperFoods Rx* will show you how easy this is to do.

Micronutrients:
The Keys to Super Health

Do you know what a "healthy diet" consists of? Fruits? Vegetables? Low fat? Lean protein? This advice is okay as far as it goes, but given what we now know about the relative nutritional values of foods, these vague guidelines are only a part of a larger picture. Many people who believe they are eating a "good" diet would be shocked at how poor their nutritional status actually is. It's a paradox that nutritional deficiencies are common in the overfed. Many people, even those of us who are eating a "healthy diet," are deficient in many of the nutrients that could be helping us to prevent disease.

SuperFoods Rx is based on the premise that we must develop a more sophisticated appreciation of the familiar building blocks of diet—the macronutrients of fat, carbohydrates, and protein—and then move on to an examination of the micronutrients in foods. All foods are not created equal. We're familiar with the idea that some proteins are better than others. Striped bass, for example, is better for you than a fatty pork chop. Many of us know that low-fat or nonfat dairy foods are better for us than full-fat ones. But the idea that one vegetable or fruit might be better than another is entirely new. We've only been able to make these kinds of distinctions because we can now examine the micronutrients in fruits and vegetables and assess which ones have more health-promoting qualities.

Micronutrients include two categories we're all familiar with: vitamins and minerals. The most exciting category of micronutrients and one that you'll hear more and more about in the coming years is phytonutrients. Phytonutrients ("phyto," from the Greek word for plant) are naturally occurring substances that are powerful promoters of human health. *SuperFoods Rx* provides specific information on phytonutrients and their effect on your health and daily energy.

INTRODUCING PHYTONUTRIENTS

Phytonutrients are nonvitamin, nonmineral components of foods that have significant health benefits. There are literally thousands of them in our foods, appearing in everything from our cup of morning tea to a handful of popcorn at the movies. Some phytonutrients help facilitate the ability of our cells to communicate with one another. Some have anti-inflammatory abilities. Some help prevent mutations at the cellular level. Some prevent the proliferation of cancer cells. Some have functions that we are only beginning to understand, and many have yet to even be identified.

Here are just three important types of beneficial phytonutrients:

Polyphenols act as antioxidants, have anti-inflammatory properties, and are antiallergenic, among other health-promoting abilities. Some foods that contain polyphenols are tea, nuts, and berries.

Carotenoids are the pigments found in red and yellow vegetables—think tomatoes, pumpkin, carrots, apricots, mangoes, sweet potatoes. They are an important category of phytonutrients that includes beta-carotene, lutein, and lycopene. These nutrients function as antioxidants; they protect us from cancer and help defy the effects of aging.

Phytoestrogens, literally "plant estrogens," are naturally occurring chemicals found particularly in soy foods as well as in whole wheat, seeds, grains, and some vegetables and fruits. They play a role in hormone-related cancers such as prostate and breast cancers.

HOW MICRONUTRIENTS CAN PROLONG YOUR HEALTH SPAN

Your body is a complicated, interrelated system that is remarkably resilient. Nonetheless, over a lifetime, the tiny links in the chain that your health depends upon begin to break down. The micronutrients in whole foods provide the reinforcements that retard this breakdown. One critically important function of micronutrients in maintaining your health is their activity as powerful antioxidants. Just as a bicycle frame in the back of the garage will eventually begin to rust, so our bodies at the cellular level "rust," or oxidize. This oxidation creates long- and short-term health problems. Antioxidants protect the body from oxidation. The antioxidants that have been the most studied and have received the most attention include vitamin C, vitamin E, beta-carotene, and minerals such as selenium. You can see the antioxidant activity of vitamin C in your own kitchen: a slice of apple will begin to turn brown shortly after being cut, but if it's rubbed with lemon juice (high in vitamin C) it will be preserved. The vitamin C slows down the oxidation process. The list of antioxidant nutrients grows almost daily. Here's a brief synopsis of how antioxidants help preserve our health.

Our bodies are heat-generating machines that depend on oxygen to carry out basic metabolic functions. One of the by-products of this use of oxygen, or "oxidation," is oxygen molecules that have been transformed into what are known as "free radicals." Free radicals are generated by the body's own metabolic systems. In addition, the environment is teeming with them in the form of cigarette smoke, pollution, certain foods, and chemicals. Even your drinking water and the sun that warms your face on an April morning are creating free radicals.

These free radicals, which are constantly proliferating throughout our bodies, are missing an electron. This makes them highly unstable. Driven to restore the missing electron, they seek out replacement molecules from whatever neighboring cells they can attack. Sometimes their targets are DNA, sometimes enzymes, sometimes important proteins in neighboring cells, and sometimes they attack the cell membrane itself. It's been estimated that each cell experiences ten thousand free-radical hits each day.

Clearly, no living being could survive for long without some powerful system of defense against free radicals. Antioxidants are the foot soldiers in the battle to disarm free radicals in our bodies. They neutralize free radicals, and, in effect, minimize their threat by giving up an electron in an effort to stabilize them. Stabilized, the free radicals are no longer a threat to cellular health.

Our bodies produce many antioxidants on their own, but the antioxidants in foods play a critical role in keeping free radicals in check. Indeed, it's the antioxidants in foods that inspired the once shocking but now commonly held belief in the medical community that certain foods promote health beyond their ability simply to nourish the body.

Scientists now believe that successfully combating free radicals, and the damage they instigate, is one of the keys to long-term health. In other words, we now know that it's no longer just genetics or medical advances that are responsible for your longevity and your ability to avoid chronic disease: it's your body's ability to handle free radicals. Unchecked free-radical activity has been conclusively linked to heart disease, cancer, diabetes, arthritis, vision problems, Alzheimer's disease, and premature aging.

It's this idea—that the body benefits immeasurably from a constant rich infusion of phytonutrients, as well as all macro- and other micronutrients—that is one of the cornerstones of *SuperFoods Rx*, and identifying the richest sources of micronutrients in foods is one of its important features.

The Four Principles of SuperFoods Rx

***SuperFoods Rx* presents a very simple idea** that rests on a cluster of important principles. Understanding these principles will help you shift the focus of your diet and thereby improve your short- and long-term health.

PRINCIPLE ONE: SUPERFOODS RX IS THE "BEST DIET IN THE WORLD"

The first question most people have about the *SuperFoods Rx* diet is what makes one food more super than another? How were the foods chosen?

As you might imagine, choosing one food over another is not a simple matter. The guiding principle is which food, within a given category, is at the top of its class in promoting health. Also, I had to consider which foods had the most desirable nutrient density, in other words, the most known beneficial nutrients and the least negative properties like saturated fat and sodium.

Today's sophisticated computers have enabled researchers to determine which human populations are the healthiest and live the longest. These epidemiological studies have also allowed us to discover the particular foods eaten by those healthy populations. Certain foods pop up over and over again when you look at the diets of the healthiest people in the world. For example, the traditional Greek diet prior to 1960, including the traditional diet of Crete, is known to be one of the most healthful diets in the world. This Mediterranean diet is primarily a plant-based diet with a number of protective substances in the most popular foods such as selenium; glutathione; resveratrol; a good balance of essential fatty acids (omega-3s to omega-6s); and high amounts of fiber, folate, antioxidants, and vitamins C and E. You've probably heard of the Okinawan diet, also recognized as healthy. (Okinawa reportedly has more centenarians—people 100 and above—as a percentage of the population than any region in the world.) My approach in selecting the SuperFoods was to analyze these diets along with other healthy consumption patterns to discover the critical foods that show up over and over.

Approaching from the other direction, in an effort to see which foods were the best proven health promoters, I also studied many highly respected databases and sets of recommendations, including those of the American Heart Association, the American Cancer Society, the National Cancer Institute, and others. These groups have made very specific recommendations based on literally countless studies on what constitutes a healthy diet. The USDA (U.S. Department of Agriculture) has very useful information. There is something called an ORAC score, in which foods are ranked according to their Oxygen Radical Absorption Capacity, or how well they act as antioxidants. Spinach and kale are the two vegetables with the highest ORAC score. That's important information, but it's not the whole picture. The ORAC score, for example, doesn't count fiber, but it does give you a starting point. Multiple charts exist on the relative amounts of nutrients in foods. I have tried to use the most accurate and up-to-date data in all cases.

I turned to the researchers themselves. I've attended many research meetings where I had the opportunity to discuss the latest findings with those who are in the front lines of nutrition research. It's extremely exciting to hear a paper presented for the first time that outlines a new finding that will affect the way people think about food. The fascinating information that has emerged on fats is something that has been getting a great deal of attention lately and is frequently discussed at these meetings. We are in a dual crisis right now: we're eating too much fat, and much of it is the wrong kind of fat. Good fat is essential to life, as you'll see in the chapter on wild salmon (page 130).

It would have been impossible to do this work even a few years ago, as much of the information I can now access is brand-new. In one instance, I hired a renowned scientist to do the analyses for me. For example, the polyphenol amounts in selected, readily available brand-name 100 percent fruit juices and jams have never been published before. You'll see in the chapter on blueberries how impressive certain juices are (and, conversely, how unimpressive others are!). The *SuperFoods Rx* guiding goal is to identify the best, buy the best, and eat the best!

PRINCIPLE TWO: SUPERFOODS ARE WHOLE FOODS

This is a *very* important principle. I am not at all opposed to supplements; I take them myself. But if there's one thing you should learn from *SuperFoods Rx*, it's that nutritious whole foods must be at the center of your nutrition plan; you can't rely solely on supplements to do the job.

Whole foods are the answer. What are whole foods? While perhaps there will always be some disagreement on a precise definition of the term, in general, whole foods are those that are unprocessed or are minimally processed and in such a way that none of their nutritional characteristics have been intentionally modified. Canned tomatoes, for example, are processed. In the processing, some of the vitamin C, for example, is lost, but the processing actually increases tomatoes' nutritional value by concentrating the remaining nutrients. For the purposes of *SuperFoods Rx*, I consider canned tomatoes a whole food.

••

What About Organic Foods?

There's no question that organic foods are better for the environment and thus for all creatures (including us) in that environment because they reduce the threat of pesticides. But are they better from a nutrient standpoint? This is a developing story. There are preliminary and still very controversial data suggesting that organically grown food—particularly some fruits and vegetables—may have more vitamin C, minerals, and polyphenols than conventionally grown varieties. Confirming evidence doesn't yet exist. Until larger studies are done confirming these preliminary data, the only definite benefit to organic foods (and one that I believe is considerable) is to the environment.

••

Phytonutrient research is an emerging science. Optimal as well as safe intakes of phytonutrients have not yet been established for many of these compounds. So we are left with whole foods to provide them in a safe and much more satisfying way. Once again, it's the synergy that matters. It's not just one particular phytonutrient in a food that makes the difference; it also seems that fiber, vitamins, minerals, and other substances in that food also enhance and regulate the actions of the phytochemicals.

Whole foods are complex. They contain as-yet-unidentified compounds that can magnify the effects of identified phytonutrients. A growing body of research from laboratory and human studies suggests that these phytonutrients work best in concert. Moreover, just as the phytonutrients in a particular food team up to combat disease and enhance well-being, so do the phytonutrients from a wide range of foods work together to promote good health.

While many studies have focused on beta-carotene, for example, there is still some uncertainty about whether the benefits associated with that carotenoid are actually due to the action of beta-carotene or to one or more of the other carotenoids found in our food. Most likely, the answers lie in the synergistic effect of multiple carotenoids working together or in some compound that hasn't yet been identified. Until all the answers are in, and that surely won't happen in the near future, the safest and most effective way to benefit from the bounty of nutrients in nature's precisely calibrated form is to eat whole foods.

About 25 percent of the population is "salt sensitive," which means that their bodies are particularly sensitive to sodium in the diet, and as a result, they are particularly vulnerable to developing hypertension. Everyone tends to become more salt sensitive as they grow older. I recommend that you try to avoid as much added salt in your diet as possible. Learn to read food labels and avoid products with lots of sodium. Remember: you can always add salt to your taste at home. It's far better for you to control the amount.

PRINCIPLE THREE: SUPERFOODS RX EQUALS SYNERGY

We now know a great deal about the rich array of micronutrients in various foods, but we still only have part of the picture. We do know that food synergy is critical to health. Food synergy refers to the interaction of two or more nutrients and other healthful substances in foods that work together to achieve an effect that each is individually unable to match. For example, the power of nuts to prevent cardiovascular disease is far greater than one would assume by looking at any single nutrient they contain. Nutrients work in a precisely calibrated relationship—the kind of relationship that nature has provided when the nutrients are obtained from food. There are, after all, thousands of chemicals present in food and researchers have only identified a fraction of this number. Surely, with this many chemicals present, interactions are taking place that science doesn't fully understand. You can't shortcut your way to good nutrition; you must rely on whole foods.

We don't have the full picture on foods and how they behave in the body. Remember, too, that some foods have attracted more attention and research interest than others, so we know more about those foods. What we don't know is precisely how all the nutrients in a given food work together to promote health. Sometimes we know the result, e.g., a high level of lutein in the macula of the eye is a predictor of visual health. We may not, however, be certain about what other substances are working along with the lutein to achieve a lifetime of excellent vision.

Spinach is a good example of the synergy of multiple nutrients. Spinach is the single food that most epidemiological studies associate with the lowest levels of cancer, heart disease, cataracts, and macular degeneration. It's very clear: the more spinach people eat, the less likely they are to develop any of those diseases. It would make sense that if you could develop a pill containing the significant substances in spinach, you'd have a potent weapon against cancer. In an effort to do just this, researchers have tried to deconstruct spinach to see what makes it so effective. Here's what they have come up with: spinach contains a stunning collection of micronutrients, including lutein, zeaxanthin, beta-carotene, plant-derived omega-3 fatty acids (only a very few vegetables contain these fatty acids), the antioxidants glutathione, alpha lipoic acid (spinach is the best food source of this amazingly potent antioxidant), vitamins C and E, polyphenols, coenzyme Q10, thiamine, riboflavin, vitamin B_6, folate, vitamin K, and the minerals calcium, iron, magnesium, manganese, and zinc. It also has chlorophyll, which may be a potent anticancer substance.

Wouldn't it make sense for a pharmaceutical company to formulate a capsule version of spinach? Not really. A pill containing all these substances (in the quantity available in whole food spinach) would be hopelessly large and impossible to swallow, not to mention expensive and unprofitable. Moreover, there's no certainty that we understand the exact

SuperFoods Rx
LIFESTYLE PYRAMID

- Aerobic exercise most days (30–60 minutes)
- For those not currently exercising, begin with walking a minimum of 1 hour per week
- Resistance exercises (weight training) 2–3 times per week
- Stress-management practice (15 minutes most days)
- Hydration: Drink 8 or more 8-ounce glasses of water daily (may include tea and/or 100% fruit/vegetable juice)
- Sleep (7–8 hours for most people)

FOOD FOR THOUGHT:
- Daydream
- Enjoy friendships and laughter
- Spirituality
- Spend some time outdoors each day

FRUITS: 3–5 servings daily; include berries (fresh or frozen) most days

VEGETABLES: Unlimited (minimum 5–7); include dark, leafy greens most days

SUPPLEMENTS:
- Multivitamin and mineral supplement daily for most people
- Consider fish oil supplement (250–1,000 mg daily)

WHOLE GRAINS: 5–7 servings per day; see SuperFoods Rx List; include whole grain noodles/pasta, tortillas, breads, and cereals

PROTEIN:
Animal Protein: 1–2 servings daily of skinless poultry breast, fish* (see SuperFoods Rx List); may include 3 oz lean red meat every 10 days (*2–4 servings per week)
Vegetarian Protein: 1–3 servings daily of legumes, lentils, soy (i.e., tempeh, tofu), egg whites, eggs (1 per day maximum)

FOR HEALTHY BONES: 1–3 servings daily of non- or low-fat dairy, tofu, soy, fortified soy milk, fortified OJ, fish with bones (i.e., sardines, canned salmon), shellfish, dark leafy greens

SEASONINGS: Cook frequently with dried or fresh parsley, rosemary, oregano, turmeric, garlic, ginger, citrus zest, chives, red onion, and white onion

SAMPLE SERVINGS:
Fruits = 1 medium piece, 1 cup, ½ cup juice, 2 tablespoons raisins, 3 prunes
Vegetables = ½ cup cooked, 1 cup raw
Grains = ½ cup cooked grains/pasta; 1 slice bread
Meat and Fish = 3 ounces lean meat, poultry, or fish
Vegetarian Protein = 1 egg, 2 egg whites, 3 oz tofu or tempeh, ½ cup cooked beans or lentils
Dairy = ½ cup non- or low-fat cottage cheese, 8 oz non- or low-fat yogurt or milk
Fats = 1 oz (24) raw almonds, 14 walnut halves, 1 tablespoon oil, ⅜ avocado

ALCOHOL: If you choose to consume alcohol, we recommend: 1–3 drinks per week for women; 2–8 drinks per week for men

HEALTHY FATS: 1–2 servings daily of nuts, seeds, avocado, extra-virgin olive oil, canola oil, soybean oil, peanut oil, flaxseed oil

Up to 100 calories daily dark chocolate, butter, buckwheat honey, sweets, or refined breads and grains proportions of nutrients as they exist in spinach. We know that the synergy of micronutrients is critical, but we don't know yet how to achieve this in a man-made substance. The bottom line is that it's most likely impossible to make a supplement that fully duplicates the synergistic power of food. But there's really no need to: if you want to have all the health-promoting power of spinach, it's available right in your market.

The synergy of *SuperFoods Rx* really extends beyond diet. Not only do all the nutrients in food work together to make something greater than the whole, the organ systems in your body do the same. We now know that disease isn't usually something that strikes out of the blue and hits an isolated organ. For example, did you know that obesity is a risk factor for age-related macular degeneration (AMD)? Who would guess that being overweight would affect your vision?

While the focus of this book is certainly food, if you want to achieve optimum health, you must incorporate other lifestyle changes into your daily life. The SuperFoods Rx Lifestyle Pyramid visually summarizes the components of a healthy life. It's critically important to improve your diet, but if you don't exercise, you won't get the most benefit from optimum nutrition. Similarly, if you do nothing to control the stress in your life, even a perfect diet won't work to your best advantage. So, it's diet, exercise, positive social interaction, stress reduction, sufficient sleep, and even sufficient fluid intake that work together—that's synergy!—to maximize the benefit of each.

SuperFoods Rx Is an SPF Diet

Eating the foods recommended in *SuperFoods Rx* can actually increase the SPF (sun protection factor) of your skin. The key nutrients of lutein/zeaxanthin, beta-carotene, alpha-carotene, lycopene, vitamins C and E, folate, polyphenols, glutathione, isoflavones, omega-3 fatty acids, and coenzyme Q10 all help protect your skin from sun damage. This is extremely important because as the ozone layer has decreased, the ultraviolet exposure of all living organisms on earth has increased.

PRINCIPLE FOUR: SUPERFOODS RX ARE SIMPLE; SUPERFOODS RX ARE POSITIVE

Food is an emotional subject for many people. For some, it's a great, unmitigated pleasure. For many, it's a source of worry and confusion. There's the nutrition issue, the weight issue, the preparation issue . . . A healthy diet is the essential core of a healthy lifestyle, along with exercise, routine preventive medical care, adequate sleep, and stress reduction. We all live busy complicated lives, and any nutrition recommendations or advice that I or anyone else offer will be ignored if it is too complicated or challenging.

Fortunately, nature agrees; so choosing from a range of whole SuperFoods is simple.

The best approach to any health change is one that is positive. I believe that "diets" that forbid foods or make eating satisfying meals a challenge are counterproductive. Once most people understand the *SuperFoods Rx* principles, they feel liberated. *SuperFoods Rx* is about what you should eat, not what you shouldn't eat. It's not about what you shouldn't do.

If you eat a healthy diet of SuperFoods, there's room for a bit of most any kind of food that gives you pleasure, whether it's chocolate or bacon. The good stuff in the SuperFoods will help to mitigate any damage done by the bad stuff. Of course, you have to be sensible about this.

I attended a medical conference in Spain where I lectured on the dietary prevention of cataracts. The next morning when I went to breakfast with a few colleagues, it became clear that they wanted me to order first because they were afraid I'd disapprove of their choices. I ordered scrambled eggs topped with fresh salsa, a bowl of berries, and a glass of delicious fresh squeezed orange juice. They were shocked. I guess they expected me to eat a bit of unsweetened granola and drink some mineral water.

Food is pleasure. When you sit at the table, you're not a patient, you're a person. Eating should be a satisfying part of your life. *SuperFoods Rx* will help make it so.

SuperFoods Rx in Your Kitchen

No eating plan will work if you can't adapt it to your lifestyle. I think *SuperFoods Rx* offers the easiest, healthiest eating plan ever! To make it even easier, I've worked closely with nutritionists and a super chef—who are feeding real people every day—to come up with tips and suggestions on how to integrate SuperFoods into your busy lifestyle.

Here are some very practical considerations and some basic information that will help you get started with *SuperFoods Rx*.

SUPERFOODS RX AND PORTION CONTROL

The amounts of food many of us eat are out of control. When I tell people that they should be eating five to seven servings of vegetables a day, they are shocked, claiming they could never eat that much food. It's quite true that one couldn't eat seven servings of vegetables a day if their serving size duplicated the portions in many restaurants. The FDA (Food and Drug Administration) says that the standard serving of pasta, for example, is 1 cup. In most restaurants, pasta portions typically measure about 3 cups. That's about three servings! We'd be eating vegetables out of buckets if we extrapolated from restaurant-portion sizes. I believe that many people have been discouraged from following very good dietary guidelines because they've come to believe that a serving is a super-size amount of food. They believe that if they really ate that much, they'd gain a tremendous amount of weight.

••

A survey by the American Institute for Cancer Research found that more than 25 percent of Americans polled said that they decided how much food to eat at a single sitting on the basis of how much food they were served.

••

When it comes to fruits and vegetables, getting the optimum number of servings isn't hard at all when you understand what a serving size really is. For most fruits and vegetables, it's ½ cup.

Here is the *SuperFoods Rx* breakdown of serving sizes in various food categories:

VEGETABLES
   ½ cup cooked or raw vegetables
   1 cup raw greens
   ½ cup vegetable juice

FRUITS
   ½ cup chopped fruit
   ½ cup fruit juice
   1 medium piece of fruit
   2 tablespoons raisins, 3 prunes

VEGETARIAN PROTEIN

    1 egg or 2 egg whites

    3 ounces tofu or tempeh

    ½ cup cooked beans or lentils

NUTS

    2 tablespoons peanut butter or 1 ounce raw nuts and
      seeds

FISH AND MEAT

    3 ounces cooked lean meat, poultry, or fish

WHOLE GRAINS

    1 slice whole wheat bread

    ½ cup cooked grain or pasta

HIGH CALCIUM FOODS

    ½ cup nonfat cottage cheese

    8 ounces nonfat yogurt or milk

FATS

    1 ounce (24) almonds, 15 walnut halves

    1 tablespoon oil

    ⅜ avocado

Here is a general idea of what you should be eating on a weekly basis:

	Daily Servings
Vegetables	5 to 7; include dark leafy greens most days
Fruits	3 to 5
Soy	1 to 2
Animal protein	0 to 3

	Daily Servings
Vegetarian protein	3 to 6
Healthy fats	1 to 2
Whole grains	5 to 7
High calcium foods	2 to 3

	Weekly Servings
Nuts and seeds	5
Fish	2 to 4

• •

Here are some tips from the American Dietetic Association on how to recognize appropriate serving sizes:

A medium potato should be the size of a computer mouse.

An average bagel should be the size of a hockey puck.

A cup of fruit is the size of a baseball.

A cup of lettuce is four leaves.

Three ounces of meat is the size of a cassette tape.

Three ounces of grilled fish is the size of your checkbook.

One ounce of cheese is the size of four dice.

One teaspoon of peanut butter equals one dice.

One ounce of snack foods—pretzels, etc.—equals a large handful.

• •

UNDERSTANDING EACH SUPERFOOD

There are fourteen SuperFoods. That does not mean that any diet should be limited to these foods! Variety in food choices is absolutely critical to health. The fourteen Super-Foods are the "flagship" foods in a given category. They were chosen because of the high concentrations of nutrients or the otherwise hard-to-get nutrients they contain, as well as the fact that many of them are low in calories. You will see that under each SuperFood is a list of the primary nutrients that have elevated them to SuperFood status. This is not meant to be a complete list of every single nutrient that food contains, but rather a list of the high-profile nutrients that have demonstrated definite health benefits and are present in that food in sufficient quantity to make a difference.

Each SuperFood has "Sidekicks." These are foods that are generally in the same category as the flagship Super-Food and offer a similar nutrient profile. For example, almonds, along with sunflower seeds and pecans and a few others, are sidekicks to the SuperFood walnuts. It's best to vary your nut consumption by choosing walnuts, or any other nut you like, on some days and sunflower seeds, say, another time. Some foods like soy have few sidekicks, so while soy is available in a variety of food forms, including soymilk, soy nuts, and tofu, it's the nutrients in soy and only soy that make it a SuperFood.

At the beginning of each *SuperFood Rx* section there is a "Try to Eat" recommendation. This is a guideline on how much of that particular SuperFood you should attempt to incorporate into your diet and how often. In some instances, I recommend that you eat the food daily; in others a certain number of times a week is sufficient.

At the end of almost every *SuperFoods Rx* chapter, you'll find a quick-and-easy recipe or two that my family and I enjoy at home on a regular basis. The SuperFoods Rx Salad, for example, that you'll find in the recipe section—I never tire of it.

The most enticing recipes are those you'll find starting on page 243. These recipes were developed especially for *SuperFoods Rx* by Chef Michel Stroot of the world-renowned Golden Door. They feature SuperFoods and were prepared with optimum nutrition in mind.

The SuperFoods Rx Shopping Lists on page 324 will be extremely helpful to you as you try to fit SuperFoods into your life. It's important to learn to read the nutrition labels on foods; they're a blueprint to whether or not any given food is a plus or minus to your diet. The truth is, however, we're often on the run and it's sometimes hard to compare among many items in our search for the best. It's extremely helpful to have a shortcut to the best cereals, breads, canned foods, etc. For example, C Monster juice by Odwalla in either strawberry or citrus is about the best juice you can drink. It's loaded with vitamin C, potassium, beta-carotene and beta cryptozanthin, some iron, and lots of phytonutrients. Post Shredded Wheat 'N Bran is another excellent food choice. It's high in fiber and nutrients and low in salt and simple sugars—a great breakfast choice. My patients have told me that they find my specific food recommendations particularly helpful. I'm sure you will, too.

THE SUPERFOODS RX RECIPES

Chef Michel Stroot is the world-famous chef at one of the best-known spas in the world—the Golden Door in Escondido, California. He knows how challenging it can be to satisfy and delight demanding patrons who are used to dining in the finest five-star restaurants. Michel has done an extraordinary job in creating for this book ten days of dazzling menus that rely on the health-promoting benefits of SuperFoods.

You don't need to be a chef to reproduce these dishes in your own kitchen. Simple and healthful, they make use of supermarket-friendly ingredients with enough of a twist to make you want to rush to the kitchen.

SUPERFOODS RX IN A NUTSHELL

Everyone has a different level of interest in diet. Some follow the latest research developments and like to know the scientific basis for every claim. If that describes you, you'll enjoy *SuperFoods Rx*. It's based on up-to-the-minute, peer-reviewed research. All the substantiating data can be found in the back of the book. I've emphasized human studies whenever possible and relied on animal or laboratory models only when sufficient human research results were unavailable.

Maybe you're the kind of reader who doesn't like to wade through a lot of text. You don't have to. All you have to do is read the basic *SuperFoods Rx* principles, then glance at the food chapters, check the shopping lists in each food section, and you're ready to improve your diet and your life.

Here is an extremely abbreviated outline of the major *Super-Foods Rx* recommendations:

- Eat at least eight servings of fruits and vegetables daily.
- Think healthy fat: try to increase your intake of seafood, nuts and seeds, avocado, extra virgin olive oil, and canola oil.
- Eat one handful of nuts about five days per week.
- Eat fish two to four times a week.
- Substitute soy protein for animal protein a few times a week. Try to have one or two servings of soy daily.
- Buy bread and whole grain cereals that have at least 3 grams of fiber per serving.
- Drink green or black tea, hot or chilled, daily.
- Have some yogurt for breakfast, or in a smoothie, dip or dessert every day.
- Add phytonutrient-rich 100 percent juices and jams to your diet.
- Avoid commercial snacks and baked goods, which contain many unhealthy fats, including saturated fat, trans-fatty acids, and an overabundance of omega-6 fatty acids, and sodium.
- Eliminate soft drinks, sweetened or "diet," except as an occasional treat.

PART II

The SuperFoods

Beans

SIDEKICKS: All beans are included in this Super-Food category, though we'll discuss the most popular and readily available beans such as pinto, navy, Great Northern, lima, garbanzo (chickpeas), lentils, green beans, sugar snap peas, and green peas

TRY TO EAT: at least four ½-cup servings per week

∙∙∙

Beans contain:

- Low-fat protein
- Fiber
- B vitamins
- Iron

- Folate
- Potassium
- Magnesium
- Phytonutrients

∙∙∙

Many people have relegated beans to the back of the pantry for a few reasons: they assume that beans, while good for vegetarians and "back to the land" types, don't have much to offer the average meat-eating diner. They also figure that beans take way too long to cook. Oh, yes, and then there's the gas issue. . . .

The truth is that beans are a virtual wonder food. A delicious source of vitamin-rich, low-fat, inexpensive, versatile protein, beans deserve a place at the table for those reasons alone. But the full power of beans to lower cholesterol; combat heart disease; stabilize blood sugar; reduce obesity; relieve constipation, diverticular disease, hypertension, and type II diabetes; and lessen the risk for cancer make this ancient food an extraordinary and important addition to any diet.

• •

Legumes include fresh beans like peas, green beans, and lima beans as well as lentils, chickpeas, black beans, and the whole dried bean family.

• •

Let's get the practical objections out of the way first. While it's true that most beans take a while to cook, they don't take up much *active* cooking time. In other words, beans simmer without your having to hover over a pot. An alternative to cooking beans is to use canned beans. This actually is the most practical way to introduce beans to your diet. You can open a can of chickpeas (garbanzo beans), cannellini beans (small white beans), or black beans and just toss them on top of a salad or add them to chili.

There is a negative to relying on canned beans: many varieties and brands are too high in sodium. Look for low-salt canned beans in supermarkets and health food stores. Always put canned beans in a strainer and rinse them with cool water. This will eliminate about 40 percent of the salt.

• •

In 1992, less than one-third of Americans ate beans during any three-day period. As income levels rise, bean consumption tends to decrease.

• •

As for the complaint that beans can cause gas . . . it is true that beans can cause flatulence. This is because bacteria attack the indigestible matter that remains in the intestine. Here are some hints for reducing any discomfort associated with eating beans:

- Some people find that canned beans as well as mashed beans are less gas producing.
- If you eat beans frequently in small amounts, your body will become accustomed to them and you'll reduce any digestive problems.
- Soak the beans before cooking: rinse and pick over the beans, then boil them for two or three minutes. Turn off the heat and let them soak for a few hours. Pour off the liquid, add fresh water, and continue cooking. This boiling and soaking releases a large percentage of the indigestible carbohydrate in the beans, making them easier to digest. Even though some vitamins are lost to this method, if it allows you to enjoy beans, it's to your benefit.
- Some people find that pressure-cooking beans reduces their gas-producing qualities. It also considerably speeds the cooking process.
- Try using Beano, an enzyme product that helps reduce the gas associated with foods like beans. Put a few drops of the product on the first bites of the food. It goes to work digesting the carbohydrates that would have fed the gas-producing bacteria.

BEAN HISTORY

Beans, peas, and lentils are ancient foods. Originating primarily in Africa and Asia and the Middle East, they spread over most of the globe, carried by nomadic tribes. They have been cultivated all over the earth for thousands of years. Evidence also suggests that many beans were first grown on the American continent. In North America, most of the dried beans commonly eaten are descendants of beans cultivated in Central and South America, seven thousand years ago. Portable, tasty, highly nutritious, nonperishable, and adaptable to any cuisine, beans show up in the signature dishes from many lands. Dal from India, hummus from the Middle East, and rice and beans from Latin America all make use of the versatile bean.

Beans are also known as legumes or pulses. They're an extensive family of plants distinguished by their seed-bearing pods. Some beans, like string beans, are eaten fresh, pod and all. Other legumes or relatives of beans included here are lentils and peas. (While soybeans are also beans, because of their special nutritional characteristics they have their own chapter, on page 148. Peanuts are also, strictly speaking, legumes, but because many people think of them as nuts, we've included them in the chapter on walnuts, on page 214.) Most of the beans that we refer to in this chapter are those that are eaten, when fully mature, in their dried form.

••

Don't forget that peas as well as string beans or green beans are members of the bean family. These fresh beans are readily available, are frequently served in restaurants, and they make it easy to reach or exceed your weekly quota of bean servings.

••

LOW-FAT PROTEIN

Beans are not just for vegetarians. Long regarded as "poor man's meat" because they offer an excellent source of protein, beans lost favor while Americans binged on animal protein. But as the ailments associated with animal protein have skyrocketed—particularly heart disease, some types of cancer, and diabetes—savvy consumers are beginning to recognize the value in the humble bean.

The American Cancer Society 1996 Dietary Guidelines recommends: "choose beans as an alternative to meat." It's a simple recommendation that stands on a huge and impressive body of research associating increased risk for a wide variety of diseases with animal protein and a decreased risk for those same diseases when plant-derived protein such as that from beans is substituted for animal food sources in the diet.

Beans are one of the most healthy and most economical sources of protein available. For example, 1 cup of lentils provides 17 grams of protein with only 0.75 gram of fat. Two ounces of extra-lean trimmed sirloin steak has the same amount of protein but *six times* the fat.

Lysine is the principal amino acid deficient in a large percentage of plant protein, and most beans have a generous concentration of it. As a result, beans are an ideal complementary protein for most other vegetarian protein choices. Lysine is one of the two amino acids essential for carnitine synthesis, and carnitine is essential for efficient energy production in the mitochondria—the cellular energy factory.

••

Polyphenols occur at relatively high concentrations in legumes. High levels of these important phytonutrients are present in colored beans, e.g., black, yellow, beige, red. The beans with the highest antioxidant concentrations, highest to lowest, are:

    Broad beans/fava beans
    Pinto beans and black beans
    Lentils

••

The traditional objection to the protein from beans—that it's not a complete protein—is a somewhat old-fashioned idea. It's true that beans (with the exception of soybeans, which are a complete protein) are missing two amino acids and are therefore not complete in the sense that these acids are necessary for the body to make use of the beans' protein. However, bean protein is completed by other common foods, such as nuts, dairy, and grains or even animal protein. In fact, many popular bean dishes—rice and beans, couscous and chickpeas, lentils and barley—capitalize on this combination. Many people used to believe that it was essential to eat beans and the complementary food at the same time. We now know that eating them in the same day is sufficient. For most people who eat a varied diet, certainly those eating a SuperFood diet, the bean protein would be readily available.

Beans and the complete protein issue is an interesting example of how too often we become distracted by relatively insignificant details concerning nutrition. When we eat fast-food burgers and fries at the drop of a hat, eschewing beans because they're not a complete protein, we've lost sight of the forest as we stare at the trees. I do know it's a bit more complicated than that; many people have grown up believing that meat is an important part of most meals and this cultural influence is difficult to shed. It does deserve some rethinking, now that studies suggest that substituting bean protein for red meat will help extend our health span and aid us in avoiding a host of chronic diseases. It's not just that you don't get the negatives of saturated fat with beans; you also get the positives of all the fiber, vitamins, minerals, and phytonutrients without the fat.

Keep in mind that eating plant protein leads to less calcium loss than animal protein, a benefit to those who are vulnerable to osteoporosis. In general, as you increase your protein intake, you increase the amount of calcium lost from your bones. The acidity that occurs with eating meat increases the calcium loss compared with plant protein. Moreover, plant protein provides phytonutrients plus vitamins and minerals that are bone-friendly.

••

Beans are a good source of water-soluble vitamins, especially thiamine, riboflavin, niacin, and folacin. Canned beans are often lower in these vitamins than are dried ones. Because of the other substantial nutritional benefits, this shouldn't be a reason to avoid canned beans.

••

BEANS AND YOUR HEART

Beans are a superb heart-healthy food. One study involving dietary patterns over a twenty-five-year period examined the risk of death from coronary heart disease in more than sixteen thousand middle-aged men in the United States, Finland, the Netherlands, Italy, the former Yugoslavia, Greece, and Japan. Typical food patterns were: higher consumption of dairy products in Northern Europe; higher consumption of meat in the United States; higher consumption of vegetables, legumes, fish, and wine in Southern Europe; and higher consumption of cereals, soy products, and fish in Japan. When researchers analyzed these data in relation to the risk of death from heart disease, they found that legumes were associated with a very impressive reduction in risk.

In another study, conducted over a period of nineteen years, 9,632 men and women were followed. None of the participants had heart disease when the study began. Over the nineteen years, 1,800 cases of coronary heart disease were diagnosed. But the follow-up data revealed that those men and women who ate beans at least four times a week had a 22 percent lower risk of coronary heart disease compared with those who consumed beans less than once a week. Moreover, those who ate beans most frequently also had lower blood pressure and total cholesterol and were much less likely to be diagnosed with diabetes.

••
Consumer Action Alert

Write or e-mail food processors and ask them to provide more canned beans that are low in sodium.
••

Eating beans frequently is associated with lower cholesterol levels. This isn't simply because bean protein is substituted for animal protein that adds dietary cholesterol to the diet. There's a certain amount of confusion out there regarding cholesterol. Cholesterol is a fatlike substance manufactured by the body and is also found in foods in conjunction with fats. Cholesterol is found *only* in animal foods. Many people think that if you eat a lot of cholesterol, your blood levels of this substance will be high, but this isn't really the case. There's a wide range of variation in people's responses to dietary cholesterol. Some people are quite sensitive to cholesterol in their food. However, many of us will have a minimal response to our dietary cholesterol intake. The bottom line? It's the intake of saturated fat and trans fat that really counts. This doesn't mean you should totally disregard cholesterol intake. Since cholesterol is present only in animal fats, if you focus on the amount of saturated fat and partially hydrogenated oils in your diet and try to substitute plant-derived protein like beans, you're well on your way to reducing your blood cholesterol levels and improving your overall health.

That said, it is still a healthy goal to keep your blood cholesterol levels low. How do you achieve this? Increase the amount of beans in your diet by eating ½ cup of beans each day. We all remember hearing about the power of oat bran to lower your blood cholesterol. Beans are just as effective, it turns out, and in lower quantities. In one study, people who ate 1½ cups of cooked dried beans each day experienced similar reductions in blood cholesterol to those who consumed a cup of raw oat bran. Moreover, by combining both beans and oat bran, there was a similarly successful outcome with lower, more realistic amounts of oat bran. The most important result in this study, as far as most readers are concerned, is that even eating only about ½ cup of canned beans per day made a significant difference in both cholesterol and triglyceride blood levels.

Beans are also an excellent source of fiber. (Animal protein, by the way, provides no fiber at all.) This fiber helps to keep "bad" cholesterol levels down while helping to boost the "good" levels.

• •

Bean Fiber All-Stars

Most of us don't get nearly enough fiber in our diets. In 1909, it was estimated that fiber intake per capita was 40 grams a day; in 1980, it was about 26.7 grams a day. It's now about 15 grams a day. That's too low! Beans give you a great fiber boost. Here's the fiber content of a few beans in ½-cup servings:

Lentils	8 grams
Black beans	7.5 grams
Pinto beans	7.5 grams
Kidney beans	5.5 grams
Kidney beans (canned)	4.5 grams
Chickpeas	4 grams

• •

It's not just the cholesterol-lowering ability of beans that's good news for your heart. Beans are also a rich source of the B vitamin folate. Lentils are particularly high in both folate and fiber. Folate plays a critical role in the reduction of homocysteine levels. Without adequate folate, homocysteine levels raise. Since homocysteine is damaging to blood vessel walls, when it accumulates, it poses an increased risk of cardiovascular disease. Elevated homocysteine levels are found in between 20 to 40 percent of patients with coronary artery disease. Just 1 cup of cooked garbanzo beans provides 70.5 percent of the daily requirement for folate. Along with the folate, beans deliver a healthy dose of potassium, calcium, and magnesium, a mineral and electrolyte combination that's associated with a reduced risk of heart disease and hypertension.

BEANS AND BLOOD SUGAR

The plentiful soluble fiber in beans is a boon to your blood sugar. If you have insulin resistance, hypoglycemia, or diabetes, beans can help you balance blood sugar levels while providing steady slow-burning energy. The fiber in beans keeps blood sugar levels from rising too rapidly after a meal. Researchers compared two groups of people with type II diabetes who were fed different amounts of high-fiber foods. One group ate a diet that provided 24 grams of fiber per day. The other group ate a diet containing 50 grams of fiber a day. The higher-fiber diet resulted in lower levels of both blood sugar and insulin. The high-fiber group also reduced their total cholesterol by almost 7 percent, their triglyceride levels by 10.2 percent, and their VLDL (very low density level) by 12.5 percent.

BEANS AND OBESITY

Beans play an important role in weight management. The simple fact is that beans fill you up: they provide lots of bulk without a lot of calories. When you add beans to your diet, you're more likely to get full before you can get fat. Beans' high-fiber content controls blood sugar and thus helps to keep hunger at bay while helping to maintain energy levels.

BEANS AND CANCER

There is promising evidence that beans may help to prevent cancer, particularly pancreatic cancer and cancers of the colon, breast, and prostate. In one study, the bean consumption and cancer rate of fifteen countries was compared, and the analysis revealed that higher rates of bean consumption were associated with a decreased risk of colon, breast, and prostate cancers. Beans contain phytoestrogens called "lignins" that have been shown to have estrogenlike properties. Researchers speculate that a high consumption of foods that are rich in lignins may reduce the risk of cancers that are related to estrogen levels—particularly breast cancer. The lignins may also have a chemopreventive effect on cancers of the male reproductive system. There are other compounds in beans called "phytates," which may be able to help prevent certain types of intestinal cancer. Epidemiological studies have shown a lower rate of cancer among people who consume higher quantities of beans, and the thought is that this result is in part because of the phytates in beans.

BEANS IN THE KITCHEN

Everyone can find some beans that they like. Most people like green beans, though they often forget that they're part of the legume family. Many home gardeners grow sugar snap peas, which are sweet and delicious. They can be enjoyed raw or cooked, and many children who refuse other vegetables will enjoy raw sugar snap peas.

If you or your family is dried-bean–resistant, start your bean exploration with lentils. They're easy to prepare because they cook quickly. They're delicious, extremely nutritious, and they're less gas producing than some other beans.

••

Be sure that lard is not an ingredient listed on canned refried beans. Look for beans that are labeled "vegetarian" on the can.

••

Here are a few popular varieties of beans:

- Adzuki beans are small, russet-colored beans. They have a thin white line on the ridge. They're somewhat thick-skinned with a sweet, nutty flavor.
- Black beans or turtle beans have a beautiful matte black color. The flesh is cream-colored with a rich, earthy flavor.
- Cannellini or white beans are white and kidney-shaped. They have a creamy, smooth texture and are good in soups and salads.
- Chickpeas are round, cream-colored legumes. Extremely popular in the Mediterranean, India, and the Middle East, they have a nutlike flavor. Very high in fiber and

nutrients, they're great when tossed on a salad, mixed with some chopped onion and olive oil, or pureed into hummus.

- Fava beans or broad beans are usually available whole in their pods or peeled and split. They're large and light brown with a nutty taste and a slightly grainy texture.
- Great Northern beans are large white beans with a creamy texture. Use them in baked bean dishes.
- Navy beans are small white beans and are so called because the U.S. Navy used to keep them aboard ships as a standard provision.
- Pinto beans are perhaps the most popular beans in the United States. Pale pink with streaks of brown, once cooked, they turn entirely pink. They have a rich, meaty taste.

SHOPPING FOR BEANS

The most important thing to remember when shopping for dried beans is to find a source with a good turnover. If you have ever boiled a pot of beans for hours and hours only to find them as tough as when you began, you can testify to the importance of using fresh beans. Even though beans are dried, they shouldn't be too old or they'll never become tender when cooked. The problem is that it's hard to tell if they're old: there's no change apparent to the naked eye. Buy beans and use them soon. People get into trouble when they buy beans, put them in the back of the cupboard and, a year later, try to cook them!

If you shop for beans in a store with open bins, be sure that the bins are covered and kept clean. Check bags of beans for powder, which indicates older beans. Be sure that the beans you buy are whole and not broken.

COOKING BEANS

Before cooking beans, spread them out and pick them over, looking for tiny stones, bugs, or clumps of dirt. Rinse the beans in a strainer under cold running water. Ideally, you'll then soak the beans for an hour or even overnight. In general, the longer you soak them, the shorter the cooking time and the less gas they'll produce. Beans that have been recently harvested generally require shorter soaking and cooking times. Beans respond well to pressure-cooking, which cuts down considerably on cooking time. Most pressure cookers have accompanying cookbooks that describe the best method to use to cook beans in that model pressure cooker.

Here are some bean ideas:

- Hummus is quick and easy to make. Puree canned garbanzo beans with chopped garlic and olive oil.
- Bean salads are fast to make. Toss different varieties together for a colorful salad with some fresh herbs and olive oil.
- Baked beans count, too! Buy or make them without too much added sugar or salt.
- Don't forget about lima beans or green peas. They're available year round, frozen, in the supermarket. Baby lima beans are delectable.
- Combine beans with pasta. I like pasta e fagioli.
- Mash beans with some finely chopped garlic and use as a sandwich spread.

See the SuperFoods Rx Shopping Lists (page 330) for some recommended beans.

Blueberries

••

Blueberries contain:

- A synergy of multiple nutrients and phytonutrients
- Polyphenols (anthocyanins, ellagic acid, quercetin, catechins)
- Salicylic acid
- Carotenoids
- Fiber
- Folate
- Vitamin C
- Vitamin E
- Potassium
- Manganese
- Magnesium
- Iron
- Riboflavin
- Niacin
- Phytoestrogens
- Low calories

••

Blueberries: Now here's a SuperFood you can take to the bank! Everybody loves blueberries and there are few foods more densely packed with healthful benefits. I tell my patients that blueberries are one of the three major Super-Foods, along with spinach and salmon. If you learn nothing else from *SuperFoods Rx*, remember to eat blueberries and spinach most days and salmon, or its sidekicks, two to four times a week. These three foods alone will change your life and health.

A small but mighty nutritional force, the blueberry combines more powerful disease-fighting antioxidants than any other fruit or vegetable. As one positive report after another has come out on blueberries, the media have taken to calling them "brain berries" and "youth berries," and they certainly deserve the good press: just one serving of blueberries provides as many antioxidants as five servings of carrots, apples, broccoli, or squash. In fact, ⅔ cup of blueberries gives you the same antioxidant protection as 1,733 IU of vitamin E and more protection than 1,200 milligrams of vitamin C.

The extraordinary health and antiaging benefits of blueberries include their role in lowering your risk for cardiovascular disease and, most likely, cancer. And we can't forget their help in maintaining healthy skin and reducing the sags and bags brought on by age. One recent study, published in the *Journal of Clinical Nutrition*, found that people who ate the equivalent of 1 cup of blueberries daily had an increased level of antioxidants in their blood—an increase that is now being studied as a "physiologic state" which plays an important role in the prevention of cardiovascular disease, diabetes, senility, cancer, and degenerative eye diseases like macular degeneration and cataracts. We know that increased blood levels of antioxidants have been shown to favorably modify incidences of breast cancer. (I found this study to be particularly interesting, as it was based on the use of whole foods rather than extracts or supplements.) Perhaps the most exciting recent news in connection with blueberries and your health is the discovery that blueberries seem to reduce the effects of age-related conditions such as Alzheimer's disease and dementia. Thus far the results on the power of blueberries have been based primarily on animal studies, but if the human clinical studies that are under way go half as well, it will be one of the most important advances in recent medicine and nutrition.

THE ALL-AMERICAN BLUEBERRY

Blueberries are native to North America. Long recognized as nutritional powerhouses, blueberries were an important part of the American Indians' diet. Originally called star fruit because of the star shape at the blossom end of each berry, blueberries were used, among other things, as a preservative. Because of their high levels of antioxidants, the berries, when pounded into dried meat, slowed the rate of spoilage of the food. Early settlers learned from the Indians to use blueberries for medicinal purposes: they brewed the berries as well as the entire plant to make medicines to treat diarrhea and to ease the discomfort of childbirth.

• •

I think of the polyphenols in berries as the choir directors. The other nutrients are all members of a huge, effective choir, working together to create something much more powerful than each individual voice. With that in mind, and remembering that each polyphenol in each berry has something to contribute, mix it up! Don't limit your berry consumption to any particular kind. Eat them all!

• •

Blueberries (called "bilberries" in Europe) work their magic primarily because of their incredibly high levels of antioxidant phytonutrients—particularly one type in the flavonoid family called "anthocyanin." Anthocyanin pigments give blueberries their intense blue-purple color. Indeed, the darker the berry, the higher the anthocyanin content. Blueberries, particularly wild blueberries, have at least five different anthocyanins. The anthocyanins are concentrated in the skin of berries because, as with many other fruits and vegetables, the plant skin protects the fruit from the sun and other environmental assaults by concentrating antioxidants in this key location.

Anthocyanins are one of the phytonutrients that give blueberries their powerful antioxidant and anti-inflammatory abilities. As we know, free radicals are the culprits that damage cell membranes and DNA and ultimately cause many of the degenerative diseases that plague us as we age. Anthocyanins are key players in neutralizing free-radical damage to cells and tissues that can lead to a multitude of ailments. The anthocyanins also work synergistically with vitamin C and other key antioxidants. They strengthen the capillary system by promoting the production of quality collagen—the building block of tissues. This important subclass of flavonoids also promotes vasodilation and has an inhibitory effect on platelet aggregation, an aspirinlike effect on blood clot formation.

••

Berries, especially cranberries, are a rich source of the flavonoid quercetin, which has been shown to possess significant anti-inflammatory properties. A recent article on the flavonoids (a class of polyphenols in berries and grapes) concluded, "Although there is still much to be learned, there are indications that the scientific approach may reaffirm the basis for many of the remedies known from traditional therapeutic use of grapes and berry products in folk medicine."

••

THE BENEFITS OF BERRIES

The health benefits of blueberries are truly impressive. For many years, researchers paid little attention to the fruit because they knew that its vitamin C levels were relatively low compared with other fruits and it didn't seem to offer any other impressive benefits. But gradually, as the power of antioxidants and in particular flavonoids—a class of polyphenols—was discovered, blueberries gained more and more attention.

The research that really put blueberries on the health map because it gained so much national attention had to do with the exciting news that the berries seemed to slow and even *reverse* many of the degenerative diseases associated with an aging brain. As we're now facing a ballooning population of aging adults—by the year 2050 more than 30 percent of Americans will be over age 65—any positive news that relates to preventing degenerative diseases like Alzheimer's or dementia is greeted with tremendous enthusiasm.

This particular berry/brain research was conducted at the USDA Human Nutrition Research Center on Aging, at Tufts University. Dr. James Joseph, director of the study, supplemented the diets of aging rats (comparable to a 65- to 70-year-old human) with the equivalent of ½ to 1 cup of blueberries, a pint of strawberries, or one large spinach salad. The blueberry-supplemented group not only performed better than the others on the various rat brain-teasers, they also showed actual *improvements* in coordination and balance. This was very impressive news indeed, as previously it was thought that degeneration due to aging was virtually irreversible. Dr. Joseph's continuing research has confirmed that blueberries have a functional antioxidant and anti-inflammatory effect on brain and muscle tissue.

How exactly did the blueberries—in an amount that would be the equivalent of a human serving of 1 cup of blueberries a day—accomplish this dramatic improvement? Three factors seem to distinguish the blueberry-fed rats: their brain cells seemed to communicate better, their brains seemed to have fewer damaged proteins than would be expected and, finally and most encouraging, their brains actually developed new brain cells. Studies are currently under way to see if these extremely impressive results can be duplicated in humans. Preliminary studies show that people who consumed a cup of blueberries daily have performed 5 to 6 percent better on tests of motor skills than a control group. We also know that there has been a positive effect on people with multiple sclerosis. This isn't surprising because the nutrients in blueberries have an affinity for the areas of the brain that control movement.

••

My kids tease me about the mountains of jam I put on my toast and pancakes. I tell them that jam consumption has been inversely associated with skin wrinkling. So I'm actually doing my face two favors: the smile I have when I eat the jam and the smooth skin I'll have in the future.

••

Here's a breakdown of some of the polyphenol contents of various juices and jams found in American markets. Some of these analyses are published here for the first time; all the juice and jam data are from independent research. Optimum daily amounts of this class of phytonutrient have not yet been determined.

Juices	milligrams of polyphenols per 8-ounce serving
Odwalla C Monster	845

Juices	*milligrams of polyphenols per 8-ounce serving*
Trader Joe's 100% Unfiltered Concord Grape Juice	670
R.W. Knudsen 100% Pomegranate Juice	639
R.W. Knudsen 100% Cranberry Juice	587
R.W. Knudsen Just Blueberry	425
L & A Black Cherry Juice	345
27% cranberry juice cocktail	137
100% apple juice	61

Jams	*milligrams of polyphenols per 20 grams—a trace over 1 tablespoon*
Trader Joe's Organic Blueberry Fruit Spread	400
Knott's Pure Boysenberry Preserves	300
Trader Joe's Organic BlackberryFruit Spread	280
Trader Joe's Organic Strawberry Fruit Spread	120
Trader Joe's Organic Morello Cherry Fruit Spread	120
Sorrell Ridge Wild Blueberry Spreadable Fruit	100
Knott's Bing Cherry Pure Preserves	100
Welch's Concord Grape Jam	60

While the brain research is perhaps the newest and most positive news about the power of blueberries, there are other equally impressive data on their health-promoting abilities. In addition to the brain-boosting anthocyanins, blueberries provide another antioxidant known as ellagic acid. Research suggests this antioxidant blocks the metabolic pathways that can promote cancer. Various studies have demonstrated that people who consume fruits with the most ellagic acid were three times less likely to develop cancer than those who consumed little or no dietary ellagic acid. Ellagic acid is found in black and red raspberries, boysenberries, Marionberries, and blackberries. This phytonutrient tends to be concentrated in the seeds (berry seeds are loaded with bioactive components). The aforementioned berries have three to nine times as much ellagic acid as three other good sources—walnuts, strawberries, pecans—and as much as fifteen times the ellagic acid found in other fruits and nuts.

••

While the polyphenol amounts in certain juices can be quite high, the calorie count can be high, too. Whole fresh fruit will always have the lowest number of calories. Don't become so enthusiastic about polyphenols that you sip juice all day and develop a calorie overload. Get your polyphenol boost by mixing half water or seltzer and half 100 percent juice.

••

American Indians were right about the blueberry's ability to promote digestive health: rich in pectin, a soluble fiber, blueberries work to relieve both diarrhea and constipation. Moreover, the tannins in blueberries reduce inflammation in the digestive system, and polyphenols have also been shown to have antibacterial properties.

Like cranberries, blueberries are a plus for urinary-tract health. Components in blueberries reduce the ability of *E.*

coli, a bacterium that commonly causes urinary tract infections, to adhere to the mucosal lining of the urethra and bladder.

••

Commercially produced berry-grape-pomegranate juice can be very rich in anthocyanins. Commercial pomegranate juices, for example, show antioxidant activity three times higher than red wine and green tea. This is because the processing extracts some of the tannins in the rind. To find the best juices, look for those with sediment at the bottom of the bottle. This indicates bits of skin, which are the prime sources of the beneficial berry-grape-pomegranate antioxidants. Shake before serving.

••

THE FRENCH PARADOX

You've probably heard of the French Paradox. It refers to the seeming contradiction discovered in regions of France where, despite a high intake of dairy fat, the people had low incidences of cardiovascular disease. At first, it was believed that the alcohol in the wine people drank was the factor that helped reduce their risk. As time went on, it was discovered that the paradox is only partly explained by the ability of alcohol to increase HDL, or "good," cholesterol. Recent research has concentrated on the ability of the flavonoids in wine to play an active role in reducing the risk of coronary artery disease. The extremely high level of polyphenols in red wine, which is about twenty to fifty times higher than white wine, is due to the incorporation of the grape skins in the fermenting process. The polyphenols in grape skins are known to prevent the oxidation of LDL cholesterol, a critical event in the process of the development of coronary artery disease. As James Joseph at Tufts University, who did the original blueberry research, says,

"What's good for your heart is good for your brain." Researchers have also noted a decreased risk of age-related macular degeneration with the consumption of limited amounts of red wine.

Health professionals are always cautious about recommending the consumption of alcohol because of its close association as a risk factor in other diseases. Nonetheless, for men who consume only a glass of red wine with dinner or women who consume a half glass, the health benefits are positive.

••

I drink juice every single day. I start the day with a sip of high-polyphenol juice, I drink some green tea in the midmorning, and I drink juice with dinner. Have juice—make sure it's 100 percent juice—or wine with dinner because the polyphenols in those beverages help to neutralize the adverse effects of the oxidized oils and fats in foods like the char on grilled foods. Straight fruit juice can be too sweet, so mix a couple of ounces of juice with seltzer or plain water and garnish with a lemon or lime slice. Juices can be high in calories, so don't forget to figure that into your lifestyle and compensate with more exercise or taking in fewer calories from other sources.

••

If you want to enjoy most of the benefits of a moderate consumption of red wine without the alcohol, drink purple grape juice, 100 percent dark cherry juice, 100 percent pomegranate juice, 100 percent cranberry juice, or 100 percent prune juice with added lutein, or alcohol-free red wine. Both purple grape juice and pomegranate juice have been shown to increase the antioxidants in your system. There's nothing more refreshing than adding a splash of grape juice or pomegranate juice and a slice of lemon in a glass of sparkling water.

DRIED FRUIT

Dried fruit is a powerful source of health-promoting nutrients, as the fruits' benefits remain and are actually concentrated if you measure by volume (except there's little vitamin C in dried fruit). In addition to the usual raisins, dates, and prunes, dried blueberries, cranberries, cherries, currants, apricots, and figs are now more readily available. Some fruit is heavily sprayed with chemicals to prevent pests and mold, and when the fruit is dried, the chemicals are concentrated, too. Blueberries and cranberries are not a heavily treated crop, but strawberries and grapes (and thus raisins) are, so I buy organic dried fruit when possible.

Dried fruit seems to possess significant antiwrinkle properties.

BERRIES IN THE KITCHEN

Berries are available fresh, dried, or frozen, so they can be eaten year round. My wife and I munch on fresh berries while having our morning hot beverage. My favorite way to eat them is to take a bowl of berries, add a sliced banana, pour over ½ to 1 cup of soymilk, drizzle the whole thing with 1 to 2 teaspoons of buckwheat honey, and mash it all with a fork. Sound weird? Try it; you'll be a convert.

Fortunately, as the health benefits of blueberries have come to the attention of the public, growers and producers have worked to make their fruit more widely available. You can now find frozen wild and cultivated blueberries in most every supermarket. It's not difficult to find organic berries, too. More and more 100 percent fruit juices (blueberry, cherry, pomegranate, cranberry, grape) can be found on store shelves.

··

Keep dried blueberries and cranberries on hand. They are a great addition to oatmeal along with raisins, prunes, and other dried fruit. Add them in the last minute or two of cooking.

··

Fresh blueberries in season may be cultivated or wild. The cultivated blueberries are more widely available; the wild ones only grow in the cool climates of the northern United States and Canada and are usually available at roadside stands or local markets, though you can also find them frozen in many supermarkets. Cultivated blueberries are plump with a deep blue color and a pale protective "bloom" that protects the berries from spoiling. Shake the container before buying: if they don't all move freely it could be because some are moldy or crushed. Wild blueberries are smaller with more intense flavor. By the way, ounce for ounce, you'll usually get more antioxidants in the wild blueberries, as their small size means you're getting more skin per ounce and, of course, the skin is where the health-promoting goodies are.

··

When adding fresh blueberries to batter for baking, dust them first with flour and they'll be less likely to fall to the bottom of the baking pan.

··

Fresh blueberries are delicate and deserve care. They should be washed briefly but only just before you're going to use them. If you need to store them in the fridge, be sure to pick out and discard any moldy or crushed fruit first. They will store in the fridge for a day or two in a container that allows air to circulate. They'll freeze well, but to prevent them sticking together, scatter them on a cookie sheet and put them in the freezer. Don't wash them before freezing. Once frozen, they can be put into airtight bags for storage.

Frozen blueberries greatly expand the fruit's possibilities, and it's important to know that all the animal studies on blueberries and black raspberries have been done with freeze-dried berries. I always keep at least one bag of frozen berries at home, ready to put in yogurt, mix into pancakes or muffins, or blend into a smoothie.

••

My Favorite Ways to Eat Blueberries

I'm lucky enough to be able to pick ripe berries in my organic garden, so I enjoy them in season.

Sprinkle berries and wheat germ on yogurt.

Mix frozen berries into hot oatmeal.

Toss onto cold cereal.

Whip into a smoothie with yogurt, banana, ice, and soy or nonfat milk.

Drop some onto whole wheat buttermilk pancakes just before turning them.

Enjoy a cup of berries in soymilk sweetened with buck wheat honey.

Nibble from a big bowl of fresh blueberries while sitting on the porch.

••

FRESH CRANBERRY-ORANGE RELISH

MAKES ABOUT 3 CUPS

Steer clear of the canned cranberry sauce and make your own when serving turkey or chicken. This is the recipe that appears on the bags of Ocean Spray fresh cranberries that are abundant in the supermarket in the fall. (Buy a couple of extra bags and throw them in the freezer; they keep for a long time. Use them in muffins, pumpkin bread, and pancakes, or toss them into oatmeal.)

> One 12-ounce package Ocean Spray fresh or frozen
> cranberries, rinsed and drained
> 1 unpeeled orange, cut into eighths and seeded
> ¾ cup sugar

Place half the cranberries and half the orange pieces in a food processor and process until the mixture is evenly chopped. Transfer to a bowl. Repeat with the remaining cranberries and orange slices. Stir in the sugar. Store in the refrigerator or freezer until ready to serve.

FROZEN YOGURT BLUEBERRY POPS

MAKES 12 POPS

For kids of all ages.

12 paper or foil baking cups, 2½-inch size

Zest and juices of 1 small lemon
2 cups plain nonfat yogurt
¼ to ½ cup sugar
1 pint blueberries

12 Popsicle sticks

Line twelve 2½-inch muffin pan cups with fluted paper baking cups. In a bowl, blend the lemon zest, lemon juice, yogurt, and sugar until smooth. Stir in the blueberries. Divide the mixture among the paper-lined muffin pan cups. Freeze for 1½ hours, or until almost firm; insert a Popsicle stick in the middle of each pop. Freeze until firm, about 2 hours. For longer storage in the freezer, cover with plastic wrap. To serve, peel off the paper liners from the pops; let stand at room temperature 4 to 6 minutes to soften slightly for easier eating.

See the SuperFoods Rx Shopping Lists (pages 337–38) for some recommended berry products.

Broccoli

SIDEKICKS: Brussels sprouts, cabbage, kale, turnips, cauliflower, collards, bok choy, mustard greens, Swiss chard
TRY TO EAT: ½ to 1 cup daily

••

Broccoli contains:

- Sulforaphane
- Indoles
- Folate
- Fiber
- Calcium

- Vitamin C
- Beta-carotene
- Lutein/zeaxanthin
- Vitamin K

••

It was 1992 and then president George Bush made a daring proclamation: "I'm president of the United States and I'm not going to eat any more broccoli." The horrified gasps of nutritionists could be heard from sea to shining sea. But in the end, broccoli triumphed. Perhaps in part because of the president's statement, the press took up the cause of broccoli, and anyone who'd doubted its power as one of our most valuable foods ultimately became a believer.

The timing was right for broccoli: in that same year, a researcher at Johns Hopkins University announced the discovery of a compound found in broccoli that not only prevented the development of tumors by 60 percent in the studied group, it also reduced the size by 75 percent of tumors that did develop. Broccoli is now one of the best-selling vegetables in the United States.

Indeed, broccoli and its cruciferous sidekicks are among the most powerful weapons in our dietary arsenal against cancer. That alone would elevate it to the status of a Super-Food. In addition, broccoli also boosts the immune system, lowers the incidence of cataracts, supports cardiovascular health, builds bones, and fights birth defects. Broccoli is one of the most nutrient-dense foods known; it offers an incredibly high level of nutrition for a very low caloric cost. Of the ten most common vegetables eaten in the United States, broccoli is a clear winner in terms of total polyphenol content; it's got more polyphenols than all other popular choices; only beets and red onions have more polyphenols per serving.

••

Broccoli is an excellent source of vegetarian iron.

••

Broccoli is a member of the *Brassica* or cruciferous family of vegetables. "Cruciferous" comes from the Latin root *crucifer*, meaning bearing a cross, which refers to the cross-shaped flowers of vegetables in this family. The name "broccoli" is derived from the Latin word *brachium*, meaning arm or branch, which describes the stalks of broccoli topped by a head of florets. Originally found growing wild along the coast of the Mediterranean, broccoli was cultivated by the Romans, enthusiastically adopted by Italians, and now is available worldwide. Italian immigrants brought broccoli to America. Broccoli is prized because of its delightful taste as well as the variety of textures it offers, from its flowery heads to the smooth and fibrous stalks. Most of the broccoli found today in our supermarkets comes from California. The fact that broccoli consumption doubled in the last decade of the twentieth century, following news of its cancer-fighting abilities, is encouraging. And the news on broccoli and its health-promoting abilities has become more impressive in the intervening years.

....

Raw vs. Cooked

Raw and cooked crucifers provide different anticancer phytonutrients. The raw vegetable has more vitamin C, but cooking makes the carotenoids more bioavailable.

Eat these vegetables both raw and cooked to get maximum cancer protection and health benefits. I like raw broccoli florets with a low-fat dip and raw shredded red cabbage combined with spinach in salads.

....

BROCCOLI AND CANCER

The development of cancer in the human body is a long-term event that begins at the cellular level with an abnormality that typically only ten to twenty years later is diagnosed as cancer. While research continues at a furious pace to find ways to cure this deadly killer—after heart disease the greatest killer of Americans—most scientists have come to recognize that cancer might well be more easily prevented than cured.

Diet is the best tool we all have at hand to protect ourselves from developing cancer. We know that a typical Western diet plays a major role in the development of cancers and we know that at least 30 percent of all cancers are believed to have a dietary component. Population studies first pointed to the role that broccoli and other cruciferous vegetables might play in cancer prevention. One ten-year study, published by the Harvard School of Public Health, of 47,909 men showed an inverse relationship between the consumption of cruciferous vegetables and the development of bladder cancer. Broccoli and cabbage seemed to provide the greatest protection. Countless studies have confirmed these findings. As long ago as 1982, the National Research Council on Diet, Nutrition, and Cancer found "there is sufficient epidemiological evidence to suggest that consumption of cruciferous vegetables is associated with a reduction in cancer."

A very recent meta-analysis, which reviewed the results of eighty-seven case-controlled studies, confirmed once again that broccoli and other cruciferous vegetables lower the risk of cancer. As little as 10 grams a day of crucifers (less than ⅛ cup of chopped raw cabbage or chopped raw broccoli) can have a significant effect on your risk for developing cancer. Indeed, eating broccoli or its sidekicks is like getting a natural dose of chemoprevention. One study showed that eating about two servings a day of crucifers may result in as much as a 50 percent reduction in the risk for certain types of cancers. While all crucifers seem to be effective in fighting cancer, cabbage, broccoli, and Brussels sprouts seem to be the most powerful. Just ½ cup of broccoli a day protects from a number of cancers, particularly cancers of the lung, stomach, colon, and rectum. No wonder broccoli is number one on the National Cancer Institute's list of nutrition all-stars.

The sulfur compounds in cruciferous vegetables are a
major reason these foods are such powerful chemopreventive foods. The strong smell that broccoli, cabbage, and
other cruciferous vegetables share comes from the sulfur
compounds that protect the plant as well as you. The
strong, sometimes bitter taste and smell of these vegetables
protect them from insects and animals.

The particular compounds in broccoli that are so effective against cancer include the phytochemicals, sulforaphane, and the indoles. Sulforaphane is a remarkably
potent compound that fights cancer on various fronts. It
increases the enzymes that help rid the body of carcinogens, it actually kills abnormal cells, and it helps the body
limit oxidation—the process that initiates many chronic
diseases—at the cellular level. Indoles work to combat cancer through their effect on estrogen. They block estrogen
receptors in breast cancer cells, inhibiting the growth of
estrogen-sensitive breast cancers. The most important
indole in broccoli—indole-3-carbinol, or I3C—is thought
to be an especially effective breast cancer preventive agent.
In a study at the Institute for Hormone Research, in New
York, sixty women were divided into groups, some eating a
high I3C diet containing 400 milligrams of I3C daily,
another eating a high-fiber diet, and yet a third control
group on a placebo diet. The women consuming the high
I3C diet showed significantly higher levels of a cancer-
preventive form of estrogen. The other diets showed no
increase in this substance. By the way, I3C is now available
in the form of a supplement. As this supplement hasn't yet

been widely tested, I recommend, as always, to rely first on a food source: eat some broccoli.

••

Sprouts = Super Broccoli

Researchers estimate that broccoli sprouts provide ten to one hundred times the power of mature broccoli to neutralize carcinogens. A sprinkling of broccoli sprouts in your salad or on your sandwich can do more than even a couple of spears of broccoli. This is especially good news for those few people—particularly children—who refuse to eat broccoli. Check www.broccosprouts.com to learn more about this nutrition-packed veggie and where you can buy it.

••

Broccoli has other components that help make it an all-star anticancer vegetable. We know that vitamin C plays a role in preventing cancer, and broccoli and many other crucifers are rich in this particular antioxidant vitamin. One cup of cooked broccoli contains more than 100 percent of the adult male/female RDA (recommended daily allowance) for vitamin C and 27 percent of my recommended daily dietary beta-carotene goal. Broccoli is also rich in fiber, which plays an important role in reducing cancer risk.

••

The sulforaphane in broccoli has been shown to be effective against *Helicobacter pylori*, a bacterium that is a common cause of gastric ulcers as well as gastric cancer.

••

IT'S ALL BETTER WITH BROCCOLI

If broccoli did nothing but protect us from cancer, that would be enough, but this mighty vegetable works on other fronts as well.

Broccoli and its related crucifers are rich in folate, the B vitamin that is essential to preventing birth defects. Neural tube defects like spina bifida have been linked to folic acid deficiency in pregnancy. A single cup of raw, chopped broccoli provides more than 50 milligrams of folate (the plant form of folic acid). Folate also is active in helping to remove homocysteine from the circulatory system; high levels of homocysteine are associated with cardiovascular disease. Folate also plays a role in cancer prevention. Interestingly, folic acid deficiency may be the most common vitamin deficiency in the world.

We all know how common cataracts are in our aging population. Broccoli to the rescue! Broccoli is rich in the powerful phytochemical carotenoid antioxidants lutein and zeaxanthin (as well as vitamin C). Both of these carotenoids are concentrated in the lens and retina of the eye. A single cup of raw, chopped broccoli provides 1.5 milligrams of lutein and zeaxanthin—8 percent of the *SuperFoods Rx* goal of 12 milligrams daily. One study found that people who ate broccoli more than twice weekly had a 23 percent lower risk of cataracts when compared to those who ate broccoli less than once a month. Lutein/zeaxanthin and vitamin C also serve to protect the eyes from the free-radical damage done to the eyes by ultraviolet light.

Broccoli and cruciferous vegetables are bone builders. One cup of raw broccoli provides 41 milligrams of calcium along with 79 milligrams of vitamin C, which promotes the absorption of calcium. While this is not a huge amount of calcium, it's at a low cost of calories and with the benefit of the many other nutrients in broccoli. Whole milk and other full-fat dairy products, long touted as the main sources of calcium, contain no vitamin C and are often loaded with saturated fat and many more calories than the 25 in 1 cup of raw, chopped broccoli. Broccoli also supplies a significant portion of vitamin K, which is important for blood clotting, and also contributes to bone health.

Broccoli is a great source of the flavonoids, carotenoids, vitamin C, folate, and potassium that help prevent heart disease. It also provides generous amounts of fiber, vitamin E, and vitamin B_6, which promote cardiovascular health. Broccoli is one of the few vegetables, along with spinach, that are relatively high in coenzyme Q_{10} (CoQ_{10}), a fat-soluble antioxidant that is a major contributor to the production of energy in our bodies. At least in people with diagnosed heart disease, CoQ_{10} may play a cardioprotective role.

• •

About 25 percent of the population inherit an aversion to the bitter taste of cruciferous vegetables. If this describes you, add salt, since that makes them taste sweeter. Use them in a stir-fry with low-sodium soy sauce or add them to casseroles and lasagnes.

• •

BROCCOLI IN THE KITCHEN

The good news about broccoli is that it's one of our most popular vegetables; the bad news is that we're not eating enough of it. In one study, only 3 percent of Americans surveyed reported eating broccoli in the prior twenty-four-hour period. What vegetables and fruits are we eating instead? Iceberg lettuce, tomatoes, french fries, bananas, and oranges. This isn't good! While tomatoes and oranges as well as bananas are good for you, iceberg lettuce and white potatoes, often in the form of french fries, are the top choices of many Americans when it comes to vegetables. We need to have a seismic shift in our vegetable choices. I'll give you some ideas here on how to get more broccoli and other cruciferous vegetables into your life.

One of the excellent features of broccoli is its ready availability. It's in season October through May, but it's easy to find in supermarkets all year long. While broccoli is probably most nutritious when bought at a roadside stand from an organic farmer, even frozen broccoli packs a valuable nutritional wallop. If buying it fresh, it pays to look for young broccoli: older broccoli can be tough and can also have a strong odor. Broccoli comes in a variety of green shades, from a rich sage green to deep forest green. You can even find broccoli in shades of purple. In the broccoli family are broccolini—a combination of broccoli and kale—and broccoflower—a combination of broccoli and cauliflower.

· ·

Because manufacturers usually remove most of the stalk when preparing broccoli for freezing, the remaining florets dominate the portion. Because the carotenoids, as well as other nutrients, are concentrated in the florets, this means that you can get up to 35 percent more of certain nutrients per portion from frozen broccoli than from fresh broccoli. Broccoli leaves have even more carotenoids than the florets.

· ·

When shopping for broccoli, choose tight, deeply colored, and dense florets, or flowers. (The deeper the color, the more phytonutrients!) Usually, the smaller the head, the better the flavor. Yellowing florets are signs that the broccoli is past its prime. If there are still leaves on the stalks, they should be firm and fresh-looking; wilted leaves are also a sign of an aged vegetable. Broccoli will keep in the fridge in a crisper for five to seven days. Never wash the broccoli before storing, as it can develop mold when damp.

Wash fresh broccoli thoroughly before using, soaking it in cold water if the florets seem to have sand or dirt in them. Don't discard the leaves; they're rich in nutrients. Cut off any tough part of the stalk, slicing up a few inches of the remaining stalk to hasten its cooking time, since the florets cook faster. Steaming or microwaving broccoli in very little water is the best way to cook it. Boiled broccoli can lose more than 50 percent of its vitamin C.

Some quick ways to get cruciferous vegetables on the table:

- Keep fresh or frozen broccoli on hand to use in stir-fries.
- Puree leftover broccoli with some sautéed onions and mix with low-fat milk or soymilk and add a grind or two of nutmeg for a great fast soup.
- Toss shredded raw broccoli with red cabbage and red onion, some homemade vinaigrette, and maybe some poppy seeds, for a quick slaw.
- I snack on leftover cooked broccoli just as it comes from the fridge; it's also good with some salad dressing and toasted sesame seeds.
- Shred Brussels sprouts and stir-fry them with a bit of minced garlic, olive oil, some coarsely chopped toasted walnuts or pine nuts, and a squeeze of fresh lemon juice. Toss with pasta or enjoy as a side dish.
- Stir-fry shredded cabbage with a tablespoon of sesame oil and serve as an accompaniment to an Asian meal.
- Coat cut-up broccoli or cauliflower with a little olive oil and salt. Roast at 425°F for 20 to 30 minutes. The vegetables become sweet and intense.
- Serve raw broccoli florets with hummus dip.
- I shred red cabbage into most every salad I eat at home. You don't have to put in much; just a bit gives you a great nutritional boost.

This is one class of vegetables where too much of a good thing may have adverse affects. Broccoli contains goitrogens, naturally occurring substances that can interfere with the function of the thyroid gland. Nonetheless, 2 cups a day of cooked Brussels sprouts or broccoli are completely safe. The bottom line: consumed in moderation, this class of vegetables confers a remarkable number of health benefits.

See the SuperFoods Rx Shopping Lists (page 346) for some recommended broccoli foods.

Oats

SUPER SIDEKICKS: wheat germ and ground flax-seed
SIDEKICKS: brown rice, barley, wheat, buckwheat, rye, millet, bulgur wheat, amaranth, quinoa, triticale, kamut, yellow corn, wild rice, spelt, couscous
TRY TO EAT: 5 to 7 servings a day

••

Oats contain:

- High fiber
- Low calories
- Protein
- Magnesium
- Potassium

- Zinc
- Copper
- Manganese
- Selenium
- Thiamine

••

The humble oat made nutrition history in 1997 when the FDA allowed a label to be placed on oat foods claiming an association between consumption of a diet high in oatmeal, oat bran, or oat flour and a reduced risk for coronary heart disease—our nation's number one killer. The overall conclusion from the FDA review was that oats could lower serum cholesterol levels, especially LDLs. The FDA stated that the main active ingredient that yielded this exciting positive effect is the soluble fiber found in oats called "beta glucan." The press leaped on this news and oats, particularly oat bran, became touted as the magic bullet against cholesterol. Subsequent research showed that the cholesterol-lowering effect of oat bran was less dramatic than originally thought and the oat bran story faded away.

It's time for a renewal of interest in the power of oats. New discoveries, combined with what's been known about oats for years, have shown that their health-promoting powers are truly impressive. Oats are low in calories, high in fiber and protein. They're a rich source of magnesium, potassium, zinc, copper, manganese, selenium, thiamine, and pantothenic acid. They also contain phytonutrients such as polyphenols, phytoestrogens, lignins, protease inhibitors, and vitamin E (they're an excellent source of tocotrienols and multiple tocopherols—important members of the vitamin E family). The synergy of the nutrients in oats makes them an outstanding and formidable SuperFood. Indeed, the degree of protection against disease offered by oats and other whole grains is greater than that of any of their ingredients taken in isolation. In addition to their power to reduce disease and extend your health span, oats are a flagship SuperFood for practical reasons: they're inexpensive, readily available, and incredibly easy to incorporate into your life. Oatmeal is on virtually every menu of every restaurant serving breakfast in America, and if you only remember to eat a bowl of oats regularly, you'll be on your way to better health.

Oats are an excellent source of the complex carbohydrates that your body requires to sustain energy. They have twice as much protein as brown rice. They're also a rich source of thiamine, iron, and selenium, and contain phytonutrients that show promise as an aid to reducing heart disease and some forms of cancer.

It's the cholesterol-lowering power of oats that drew the most attention to this humble grain. The specific fiber—beta glucan—in oats is the soluble fiber that seems responsible for this benefit. Study after study has shown that in individuals with high cholesterol (above 220 mg/dl), consuming just 3 grams of soluble oat fiber per day—or roughly the amount in a bowl of oatmeal—can lower total cholesterol by 8 to 23 percent. Given that each 1 percent drop in serum cholesterol translates to a 2 percent decrease in the risk of developing heart disease, this is a significant effect.

OATS AND BLOOD SUGAR

The beneficial effect of oats on blood sugar levels was first reported in 1913. In recent years, researchers have discovered some of the mechanisms that make oats so effective. The same soluble fiber that reduces cholesterol—beta glucan—also seems to benefit those who suffer from type II diabetes. People who eat oatmeal or oat bran–rich foods experience lower spikes in their blood sugar levels than they could get with, say, white rice or white bread. The soluble fibers slow the rate at which food leaves the stomach and delays the absorption of glucose following a meal. As stabilizing blood sugar is the goal of anyone with diabetes, this is an extremely beneficial effect. One recent study in the *Journal of the American Medical Association* found a low intake of cereal fiber to be inversely associated with a risk for diabetes. The authors conclude: "These findings suggest that grains should be consumed in a minimally refined form to reduce the incidence of diabetes mellitus." This same study also looked at the role of various foods in connection with diabetes. They found a significant inverse association with cold breakfast cereals and yogurt and, not surprisingly, a significant positive association with colas, white bread, white rice, french fries, and cooked potatoes. The more you eat of the latter, the greater your risk for diabetes.

....

How Much Is a Serving?

While eating five to seven servings of whole grains daily sounds like a tremendous amount, the USDA serving size is small, so it's not all that difficult to take in an adequate amount. Look for whole grain products, which are higher in fiber. Here are some typical serving sizes:

    1 slice of bread, 1 small roll, or 1 muffin
    ½ cup cooked cereal, rice, or pasta
    5 or 6 small crackers
    1 four-inch pita
    1 small tortilla
    3 rice or popcorn cakes
    ½ hamburger roll, bagel, or English muffin
    1 serving of cold cereal (amount depends on type—
    check box label)

....

OATS' POWERFUL PHYTOCHEMICALS

In addition to the power of oat fiber, researchers have been excited to learn more about the phytonutrients in grains and how they help prevent disease. The germ and bran of oats contain a concentrated amount of phytonutrients, including caffeic acid and ferulic acid. Ferulic acid has been the focus of recent research that shows promising evidence of its ability to prevent colon cancer in animals and other experimental models. Ferulic acid has been found to be a potent antioxidant that is able to scavenge free radicals and protect against oxidative damage. It also seems to be able to inhibit the formation of certain cancer-promoting compounds.

••

Corn, one of America's favorite vegetables, is actually a grain. Corn is a unique grain, as it is a source of five carotenoids: beta-carotene, alpha-carotene, beta cryptoxanthin, and lutein/zeaxanthin.

Only yellow corn has significant amounts of these healthful carotenoids; white corn does not.

••

While oats are the flagship SuperFood of this chapter, the entire category of whole grains is an important component of a *SuperFoods Rx* diet.

An unusual feature of oats is that they have two "Super Sidekicks": ground flaxseed and wheat germ. The Super Sidekicks really belong in a very special category because they're so nutrient dense. Both offer super benefits in very small amounts. If you add just 2 tablespoons each of ground flaxseed and wheat germ each day to your cereal, you will be on your way to better health.

FLAXSEEDS

Flaxseeds are a Super Sidekick that deserve special attention because these seeds are the best plant source of omega-3 fatty acids. They're a quick, easy way to get this important nutrient into your diet. (For a more complete discussion of this crucial, and often neglected, component to our diets, see the chapter on salmon.) Flaxseeds are also a powerful source of fiber, protein, magnesium, iron, and potassium: an all-around treasure trove of nutrients. Flaxseeds are also the leading source of a class of compounds called "lignins," which are phytoestrogens, or plant estrogens. Lignins influence the balance of estrogens in the body and help protect against breast cancer.

Flaxseeds are slightly larger than sesame seeds, darker in color—they range from dark red to brown—and very shiny. You can buy them in the form of flaxseed meal, or you can buy them in seed form and grind them yourself in a coffee grinder or mini food processor. The seeds must be ground, as the nutrients are difficult to absorb from the whole seed. Since the oil in flaxseeds spoils quickly, ideally it's best to grind them as you go. Some people use a grinder, dedicated to flaxseeds, and grind them in small amounts, keeping the ground portion in the fridge in a small glass jar. I keep flaxmeal—already ground flaxseed, which you can buy in health food stores—in a plastic container in the fridge. I sprinkle 2 tablespoons of ground flaxseed a day on oatmeal, cereal, and yogurt, or use it in smoothies, pancakes, muffins, and quick breads. All you need is one to two tablespoons of ground flaxseed a day. This gives you more than the Institute of Medicine's total daily recommendation for alpha linolenic acid (ALA, or plant-derived omega-3 fatty acids). Two tablespoons of ground flaxseed is a safe amount, geared to providing optimal nutrition, and there are no data suggesting that this amount of flaxseed/ALA has any deleterious effect.

••
Super Breakfasts

A bowl of hot oatmeal with raisins or dried cranberries or blueberries, sprinkled with 2 tablespoons each of ground flaxmeal and toasted wheat germ. You really can't find a better start to your day. That's my usual winter breakfast; in the summer, it's flaxmeal, wheat germ, and berries on yogurt.
••

WHEAT GERM

I grew up eating wheat germ and using it is one of the easiest ways to increase your intake of whole grains. Wheat is one of the oldest harvested grains and was first cultivated about six thousand years ago. Wheat germ is the embryo of the wheat berry (a wheat kernel that has not been heated, milled, or polished), and it's packed with nutrition. Two tablespoons, at only 52 calories, have 4 grams of protein, 2 grams of fiber, 41 micrograms of folate, a third of the RDA (recommended daily allowance) of vitamin E, along with high levels of thiamine, manganese, selenium, vitamin B_6, and potassium together with reasonable levels of iron and zinc. Wheat germ, like flaxseed, is also one of the few sources of plant-derived omega-3 fatty acids. Just 2 tablespoons—the serving size of wheat germ—of Kretschmer toasted wheat germ have 100 milligrams of beneficial omega-3 fatty acids.

Wheat germ contains phytosterols that play a role in reducing cholesterol absorption. A recent clinical trial reported that slightly less than 6 tablespoons of wheat germ per day caused a 42.8 percent reduction in cholesterol absorption among the human volunteers in the study.

Sprinkle wheat germ on yogurt or on cold cereal or hot oatmeal. Add it into pancake and muffin mix and into quick breads. When you think that only 2 tablespoons of wheat germ can significantly boost your day's nutrition, I don't know why anyone wouldn't keep a jar of it in their fridge.

WHOLE GRAIN CONFUSION

Before I make a case for the truly impressive health-promoting abilities of whole grains, I'd like to clear up some confusion that may have unfortunately encouraged you to avoid whole grain foods in the past and/or conversely might be encouraging you to buy the wrong "whole grain" foods, which are of little nutritional value.

Few issues in the diet and nutrition wars are more confusing than carbohydrates. Low-carb diets have increased the confusion: they've drawn attention to carbohydrates, but unfortunately have oversimplified the issue of protein versus carbs. Many people have come to believe that carbs equal weight gain and are bad. Foods are now being labeled with banners that claim "no-carb" or "carb-free." Consumers trying to lose weight are being told that eating carbs will destroy any hope of weight loss. What's been lost in this battle, at least for many consumers, is the fact that, like fats and protein, not all carbs are created equal.

Carbohydrates are found in a vast array of foods, from table sugar to vegetables, beans, and whole grains. A teaspoon of sugar is a carb. So is a slice of whole grain bread. You can guess which is better for you, but you may not know precisely why. This chapter—starring oats—is going to convince you that not only are carbohydrates—*whole grain carbohydrates*—good for you, they also are absolutely critical in your quest for lifelong health.

••

How to Read a Bread/Cereal Label

There are two things to look for to ensure a healthy product:

1. The list of ingredients should *begin with the word "whole."* This applies to all baked goods, including bread, crackers, cereals, pretzels, etc.

2. Look at the "Nutrition Facts" part of the label. The fiber content should be *at least 3 grams per serving for bread and cereal.* If it's lower, put the item back.

••

Whole grains lower your risk of coronary heart disease, stroke, diabetes, obesity, diverticulosis, hypertension, certain cancers, and osteoporosis. Despite what you may hear, they do not make you fat (unless you eat way too much of them—which is hard to do!). The reason carbs have the reputation for promoting weight gain is that the vast majority of those eaten by Americans are *refined carbs* like cookies, doughnuts, breads, and cakes that are loaded with sugar and fat and often trans fats as well. Yes, they are carbohydrates, but they are a world away from the complex carbohydrates that are whole grains. Many people don't realize what a difference there is between whole and refined grains. While whole grains are actually *health-promoting*, refined grains, such as pasta, white flour, white bread, and white rice, have been associated with a variety of negative health effects, such as an increased risk of colorectal, pancreatic, and stomach cancers.

If you stick with real whole grains, I promise you, you'll get full before you get fat. (How much brown rice can you eat?) You'll also increase your health span and maybe even your lifespan.

• •

White flour was not available until 1880. In 1943, some of the nutrients that had been stripped out in processing were added back, including some of the B vitamins and iron, to white flour. In 1998, folic acid was put back in. The lost vitamin E in various forms and the phytonutrients were never returned and, given the complexity of these nutrients, probably could not effectively be added back. Get everything that's missing from refined grains: *eat whole grains!*

• •

WILL THE REAL WHOLE GRAINS PLEASE STAND UP?

Some of you may avoid all grain products because you've been led to believe that such foods are carbs and will promote weight gain. I hope by the end of this chapter you'll become whole grain enthusiasts.

Other people believe they're eating healthy whole grain foods because food labels have convinced them they're buying wisely. But consider this: only 5 percent of the grain foods in the American diet are whole grains. What are 95 percent of those refined grains going into? In some cases, products that would have you believe they're healthy and nutritious are not at all. Words like "honey wheat," "multi grain," "hearty wheat," "nutri grain," in fact, do not indicate anything about the healthfulness of the product. Foods that use such words on the labels might be nutritious, but these terms guarantee nothing.

Food producers are beginning to catch on. Scan the bread shelf at your supermarket. You'll probably see a number of new breads that have health claims. They contain higher levels of fiber than you used to find, perhaps added soy flour, and they're made of whole grains. Choosing one of those healthy breads is a simple way to boost your whole grain intake and improve your overall health profile.

WHAT IS A WHOLE GRAIN?

A whole grain, whether it's oats, barley, wheat, bulgur, or a host of others, contains every part of the grain. The three parts include:

- The bran: a health-promoting, fiber-rich outer layer that contains B vitamins, minerals, protein, and other phytochemicals.

- The endosperm: the middle layer that contains carbohydrates, proteins, and a small amount of B vitamins.
- The germ: the nutrient-packed inner layer that contains B vitamins, vitamin E, and other phytochemicals.

It's the synergy of these three components that makes whole grains life sustaining. The refined carbs described earlier have been stripped of their health-promoting parts. When grains are "refined" to make white flour or white rice, for example, the bran and the germ, and all their powerful nutrients, antioxidants, and phytonutrients are stripped away, leaving a starchy substance that is to whole grain what soda pop is to 100 percent fruit juice. They can make it into bread, but they can't make it healthy!

....

How to Get 15 Grams of Whole Grain Fiber a Day

Uncle Sam Cereal—toasted whole grain wheat flakes with crispy whole flaxseed (1 cup)	10 grams
½ cup oats	9 grams
Post Shredded Wheat 'N Bran (1¼ cups)	8 grams
2 tablespoons flaxseed	7 grams
1 slice Bran for Life bread	5 grams
¼ cup oat bran (raw—not toasted)	4 grams
2 tablespoons wheat germ (crude—not toasted)	2 grams
½ cup cooked brown rice	2 grams
½ cup cooked yellow corn	2 grams

....

WHOLE GRAINS AND YOUR HEALTH

Whole grains are essential to health. They provide fiber, vitamins, minerals, phytonutrients, and other nutrients that are simply not available in any other effectively synergistic package. All healthy diets rely on them. Despite the fact that whole grains form the base of most food pyramids, indicating that they should be a significant part of our diet, many Americans fail to eat even one whole grain serving a day! Men and women who eat whole grains have a reduced risk of twenty types of cancer, according to a 1998 review of forty observational studies, published in the journal *Nutrition and Cancer*.

Whole grains also benefit the heart, according to an analysis of data from the Iowa Women's Health Study, a nine-year study of more than 34,000 postmenopausal women. When all other factors were considered, it was found that women who ate a serving or more of whole grain foods each day had a 14 to 19 percent lower overall mortality rate than those who rarely or never ate whole grains. It really is a tragedy that we consume so few whole grains and so much refined grains. If we could shift that balance, we would all be far healthier. We've already seen how oats can lower cholesterol levels and stabilize blood sugar. The complete list of the health-promoting abilities of whole grains is quite long.

••

Vitamin E intake from food, not supplements, has been inversely related to the risk of stroke. Whole grains and nuts are the two major sources of whole food vitamin E.

••

Whole grain consumption has been linked to a reduction in the risk of strokes. In the Nurses' Health Study, among the group that never smoked, a median intake of 2.7 servings of whole grains a day was associated with a 50 percent reduction in the risk of ischemic stroke. Given that less than 8 percent of adults in the United States consume more than three servings of whole grains a day, it's clear we are missing a major opportunity. When you consider that in the United States, strokes are a leading cause of morbidity and death, with an estimated 700,000 strokes annually costing roughly forty billion dollars a year, you can see that convincing Americans to add the SuperFood oats and other whole grains to their diet is well worth the effort.

One study in the *Journal of the American Medical Association* studied young adults and found those with the highest fiber intake had the lowest diastolic blood pressure readings. Hypertension is consistently the most important risk factor for stroke. Researchers estimated that a 2-millimeter decrease in diastolic blood pressure would result in a 17 percent decrease in the prevalence of hypertension and a 15 percent reduction in risk for stroke. Whole grains form an important part of the DASH diet (Dietary Approaches to Stop Hypertension; see website http://www.nhlbi.nih.gov/health/public/heart/hbp/dash/), which has repeatedly been found to lower blood pressure.

Whole grains are also helpful in preventing coronary artery disease. In the same Nurses' Health Study mentioned earlier, women who consumed a median of 2½ whole grain servings a day experienced more than a 30 percent lowered risk of coronary artery disease.

••

Whole grains contain folate, which helps to lower serum levels of homocysteine—an independent risk factor for stroke and cardiovascular disease.

••

WELCOME TO THE WIDE WORLD OF GRAINS

Oats are the flagship SuperFood for whole grains. The reasons are simple: their health-promoting benefits are considerable and they are also readily available and easy to incorporate into your diet. But there are many other whole grains that can help you increase and vary your intake of this valuable food group. Here are a few that you might want to try. Remember, it's the synergy of the various nutrients in a variety of grains that will give you optimum nutrition.

A Host of Grains and Fiber Content	*(¼-cup servings)*
Triticale	8.7 grams
Barley	8 grams
Amaranth	7.4 grams
Wheat bran (raw or untoasted)	6.5 grams
Rye	6.2 grams
Buckwheat	4.3 grams
Wheat germ (raw or untoasted)	3.8 grams
Quinoa	2.5 grams
Wild rice	1.5 grams
Millet	1.5 grams
Brown rice	0.9 gram
Enriched white rice	0.2 gram

BUYING AND COOKING WHOLE GRAINS

As whole grains become more popular, more markets carry them. If you do buy them from open bins, be sure that the store has a good turnover so the grains are fresh. Make sure that the bins are covered and kept clean.

Store whole grains in airtight containers, in a cool place, preferably the refrigerator. Oats, for example, have more natural oil than many people realize and can become rancid if they're stored in a warm environment.

Soaking whole grains before cooking can reduce the cooking time.

Many grains improve in flavor if they're toasted before cooking. Heat them in a nonstick pan over a low heat until just fragrant and they become darker, taking care not to burn them.

Once grains are cooked, they will keep in the fridge for two to three days. They freeze well, so it's a good idea to make long-cooking grains in batches that can be frozen in portion sizes. Then they can easily be added to soups, casseroles, and salads.

Here are some tips for eating more whole grains:

- Buy only whole grain bread.
- Substitute brown rice for white rice.
- Buy whole grain crackers for snacks.
- Read your breakfast cereals labels; get rid of the refined, highly sugared ones in your pantry.
- Use whole grain tortillas and pita bread for sandwiches and wraps.
- Add some oats to stuffings, meatballs, and meat loaf.
- Try some of the "exotic" grains as side dishes such as barley or quinoa.
- Look for Japanese soba buckwheat noodles. They're good in soups or cold with a sesame dressing.

APPLE-OAT CRISP

SERVES 8 TO 10

8 large Granny Smith cooking apples, cored and sliced
    (do not peel)
1½ cups rolled oats
½ cup brown sugar
¾ cup chopped walnuts
1 teaspoon sugar
2 tablespoons Smart Balance buttery spread
3 tablespoons soymilk

Arrange the apple slices in a 13 x 8-inch baking dish. You may need more or less to fill the dish. In a separate bowl, mix all dry ingredients with a fork or pastry blender and cut the buttery spread into the mixture. Drizzle the soymilk over the top and mix. The mixture should be crumbly. Put the topping on the apple slices. Cover with aluminum foil and bake in a 350°F oven for 45 minutes. Remove the foil and bake until the apples appear bubbly at the bottom of the dish. Top with fresh or frozen yogurt if you like.

See the SuperFoods Rx Shopping Lists (pages 332–34) for some recommended oat foods.

Oranges

SIDEKICKS: lemons, white and pink grapefruit, kumquats, tangerines, limes
TRY TO EAT: 1 serving daily

....

Oranges contain:

- Vitamin C
- Fiber
- Folate
- Limonene

- Potassium
- Polyphenols
- Pectin

....

Oranges may well have been America's first "health food." Long recognized as a potent source of vitamin C, oranges are considered by most to be tasty, juicy, and perhaps too familiar. No one gets very excited about an orange in a lunchbox—but they should. The discoveries that are being made about the power of oranges to support heart health and prevent cancer, stroke, diabetes, and a host of chronic ailments should bring oranges and other citrus fruits back to center stage as crucial components in a healthy diet.

Oranges originated in Asia thousands of years ago and have become one of the most popular fruits the world over. Christopher Columbus brought orange seeds to the Caribbean Islands in the late fifteenth century, and Spanish explorers then brought oranges to Florida in the next century. About two hundred years later, in the eighteenth century, Spanish missionaries brought oranges to California. These two states remain the primary producers of oranges in the United States.

HER MAJESTY'S LIFE PRESERVER

The story of how limes and the vitamin C they contained saved Her Majesty's sailors is familiar to many people. It's an interesting tale of the power of food and the importance of getting a wide range of health-promoting nutrients in your diet. In the fifteenth and sixteenth centuries, sailors were routinely lost on long sea voyages, dying of scurvy. Countless sailors died despite the belief they were getting enough to eat. The Portuguese explorer Vasco de Gama lost nearly half his men to scurvy in the 1490s as he took his first trip around the Cape of Good Hope. It wasn't until the mid-1700s that James Lind, a British naval surgeon, discovered that a daily ration of citrus in lemons, limes, or oranges preserved the sailors' health. Thus the life of the British sailor, or "limey," was saved.

This would be quaint old news were it not for the fact that 20 to 30 percent of America's adults have marginal blood levels of vitamin C and 16 percent are reportedly deficient in the vitamin. Humans (and guinea pigs) can't manufacture vitamin C in their bodies. It's water soluble and not retained in the body, so we need a constant replenishment from dietary sources to maintain adequate cellular and blood levels. Alarmingly, a high percentage of children consume minimal amounts of vitamin C. The recommended daily allowance (RDA) for Americans is 90 milligrams a day for adult males and 75 milligrams a day for adult females. This recommendation, in my opinion, is quite low. I think that the optimal intake of dietary vitamin C is 350 milligrams or more from food. But up to a third of us consume *less* than 60 milligrams of C daily.

It's fairly shocking that, given today's abundance of food, so many of us are deficient in a vitamin that's crucial to good health. While we might not be seeing cases of scurvy, we're certainly seeing epidemics of heart disease, hypertension, and cancer. The vitamin C in citrus, along with the other valuable nutrients, can play a major role in reducing these high levels of chronic disease.

••

Remember, vitamin C is rapidly excreted from the body. Adequate daily intake is critical for optimum health.

••

The precipitous decline in vitamin C levels has occurred over the past twenty years for largely unknown reasons. One possibility is that many if not most consumers have switched from orange juice in frozen concentrate form to ready-to-drink orange juices and orange drinks. As orange juice is the primary source of vitamin C in our diet, and the frozen form of orange juice is considerably higher in vitamin C than the ready-to-drink juices and drinks, this switch could be having a dramatic effect on our overall health.

We're not getting enough vitamin C from dietary sources, and oranges and orange juice are the prime source of this beneficial nutrient. There are various potent components of whole oranges that make them star performers in the fight against aging and disease. I tell all my patients to eat and drink more citrus. A simple fact: marginal vitamin C status has been linked to an increase in many causes of mortality, especially cancer and cardiovascular disease. A single navel orange, at only 64 calories, provides 24 percent of my daily dietary vitamin C recommendation of 350 milligrams or more. (That same navel orange provides 92 percent of the adult male RDA and 110 percent of the adult female RDA.) Only a limited number of fruits and vegetables are rich in vitamin C.

Vegetables	Milligrams of Vitamin C
1 large yellow bell pepper	341
1 large red bell pepper	312
1 large orange bell pepper	238
1 large green bell pepper	132
1 cup raw chopped broccoli	79

Fruits	Milligrams of Vitamin C
1 cup fresh sliced strawberries	97
1 cup papaya cubes	87
1 navel orange	83
1 medium kiwi	70
1 cup cubed cantaloupe	59

Juices	Milligrams of Vitamin C
1 cup (8 ounces) Odwalla C Monster	350
1 cup fresh orange juice	124
1 cup orange juice from concentrate	97

For all intents and purposes, the only fruit or vegetable with a high level of vitamin C that's consistently consumed in the United States is orange juice. On average, adults with a desirable vitamin C intake consume more than one daily serving of dietary vitamin C.

THE POWER OF FLAVONOIDS

Flavonoids are a class of polyphenols found in fruits, vegetables, legumes, nuts, seeds, grains, tea, and wine. There are over five thousand flavonoids that have been identified and described in scientific literature, and we're learning more about them every day. Citrus flavonoids, which are found in the fruit's tissue, juice, pulp, and skin, are one of the reasons for the health-promoting attributes of citrus fruits and the reason that the whole fruit is so much more healthful than just the juice. Two of the flavonoids in citrus—naringin in grapefruit and hesperidin in orange—occur only rarely in other plants and are thus essentially unique to citrus. The power of citrus flavonoids is dazzling. They're antioxidant and antimutagenic. The latter refers to their ability to prevent cells from mutating and initiating one of the first steps in the development of cancer and other chronic diseases. This is accomplished by their apparent ability to absorb ultraviolet light, protect DNA, and interact with carcinogens. Citrus flavonoids have been shown to inhibit cancer cell growth, strengthen capillaries, act as anti-inflammatories, and they are antiallergenic and antimicrobial. Flavonoid intake is inversely associated with the incidence of heart attack and stroke as well as a host of other ailments.

••

Rutin, a flavonoid found in citrus (and black currants), has an anti-inflammatory effect, possesses antiviral activity, and helps protect capillaries from age-related "breakdown."

••

ORANGES AND CARDIOVASCULAR HEALTH

We are certain that an orange a day promotes cardiovascular health. The Framingham Nurses' Health Study found that drinking one daily glass of orange juice reduced the risk of stroke by 25 percent. Countless other studies have confirmed similar benefits from regular consumption of citrus. We're beginning to understand that, as with so many SuperFoods, it's the synergy of multiple foods and the variety of nutrients they contain that combine to amplify and intensify individual benefits.

For example, oranges are rich in vitamin C. They are also rich in flavonoids, such as hesperidin, that work to revive vitamin C after it has quenched a free radical. In other words, the hesperidin strengthens and amplifies the effect of vitamin C in your body. In an interesting human clinical trial, orange juice was shown to elevate HDL cholesterol ("good" cholesterol) while lowering LDL (so-called bad) cholesterol.

•••

Pectin, the dietary fiber that's so effective in helping to reduce cholesterol, is present in large amounts in the white lining of citrus fruit (it's known as albedo). An easy way to increase your pectin intake is to eat the white pith. I always eat the "white stuff" on the inside of orange or tangerine rinds, scooping up a little of the orange color as well to boost my limonene intake.

•••

The fiber in oranges is another major contributor to heart health. Citrus fruit (especially tangerines) are one of the richest sources of high-quality pectin—a type of dietary fiber. Pectin is a major component of the kind of fiber that is known to lower cholesterol. Pectin is also helpful in stabilizing blood sugar. A single orange provides 3 grams of fiber, and dietary fiber has been associated with a wide range of health benefits. About 35 percent of Americans consume their fruit only in juice form. In most cases, their health would benefit if they would add whole fruit whenever possible.

Powerful Pulp

The concentration of vitamin C in orange pulp is twice that found in the peel and ten times that found in the juice. Bottom line: eat the pulp and buy high-pulp juice.
••

Oranges also help prevent cardiovascular disease by supplying folate, also called "folacin" or "folic acid," when used in supplementary form. Folate is one of the B vitamins. The total folic acid content in the average diet has been found to be below the recommended daily allowances, and mild-to-moderate folate deficiency is common. In fact, folate deficiency is known to be among the most common vitamin deficiencies in the world. This is unfortunate, as we're learning more every day about the importance of this nutrient. We know that dietary folate can play an important role in the prevention of cardiovascular disease; it is essential for the maintenance of normal DNA and also plays a role in the prevention of colon and cervical cancers, and possibly even breast cancer.

Folate plays an important role in lowering blood concentrations of homocysteine. Homocysteine is an amino acid by-product of protein metabolism and its role in cardiovascular disease is significant. High levels of homocysteine have been implicated in cardiovascular disease and even vascular disease of the eye. In one study at Harvard, men with slightly elevated levels of homocysteine were three times more likely to have heart attacks than those with normal levels. A large U.S. government–sponsored study in 2002 found that there was an inverse risk for heart attack and stroke among people who consumed the most folate: the more folate, the lower the risk. Folate works with the other B vitamins—B_{12} and B_6 and probably betaine (a plant-derived compound that seems to lower homocysteine)—to remove homocysteine from your circulatory system. The homocysteine that is allowed to build up in your body can damage your blood vessels and ultimately precipitate a "cardiovascular event." Interestingly, there is evidence that increased folate intake can actually improve heart health in people who have already developed heart disease.

••
A low intake of vitamin C can double the risk of hip fracture.
••

ORANGES AND CANCER

Recent news from researchers has demonstrated that oranges can play a significant role in preventing cancer. We know, for example, that the Mediterranean diet, which includes a considerable amount of citrus, is associated with a low incidence of cancers of the breast, lung, pancreas, colon, rectum, and cervix. Indeed, citrus fruits have been found to contain numerous known anticancer agents—possibly more than any other food. The National Cancer Institute calls oranges a complete package of every natural anticancer inhibitor known. As you might suspect, the anticancer power of oranges is most effective when the whole fruit is eaten: it seems that the anticancer components of oranges work synergistically to amplify one another's effects. The soluble fiber, or pectin, which is so effective for heart health, is also an anticancer agent. It contains antagonists of growth factors, which in the future may be shown to have a positive effect decreasing the growth of tumors. We know that in animals pectin has been shown to inhibit the metastasis of prostate and melanoma cancers.

One particular phytonutrient has attracted attention lately as a health-promoting agent. Amazingly, we routinely throw out this most potent part of the orange. In the oil of the peel of citrus fruits is a phytonutrient known as limonene. Oranges, mandarins, lemons, and limes contain significant amounts of limonene in the peel and smaller quantities in the pulp. Limonene stimulates our antioxidant detoxification enzyme system, thus helping to stop cancer before it can even begin. (It's reassuring to know that a natural chemopreventive phytonutrient can work to prevent the process of carcinogenesis at the earliest stages.) Limonene also reduces the activity of proteins that can trigger abnormal cell growth. Limonene has blocking and suppressing actions that, at least in animals, actually cause regression of tumors. One study of people in Arizona found that those who used citrus peels in cooking reduced their risk of squamous cell carcinoma by 50 percent. We've long known that Mediterranean people suffer lower rates of certain cancers than others, and researchers now believe this can partly be ascribed to their regular consumption of citrus peel. Try the Mediterranean lemonade at the end of this chapter for a great limonene boost. And, by the way, orange juice does contain some limonene but not nearly as much as the peel. Fresh-squeezed juice has the most limonene, along with other nutrients, and orange juice pulp has 8 to 10 percent more limonene than juice with no pulp.

Vitamin C, abundantly available in oranges, also plays a role in fighting cancer. In fact, there's a relatively consistent inverse association of vitamin C with cancer of the stomach, oral cancer, and cancer of the esophagus. This makes sense, as vitamin C protects against nitrosamines, cancer-causing agents found in food that are thought to be responsible for instigating cancers of the mouth, stomach, and colon. One study of Swiss men found that those who died of any type of cancer had vitamin C concentrations about 10 percent lower than those who died from other causes.

CITRUS AND STROKE

Citrus seems to have a protective ability against stroke. In the Men's Health Professionals Follow-Up Study, citrus and citrus juice were major contributors to the stroke-risk reduction from fruits and vegetables. It has been estimated that drinking one glass of orange juice daily may lower the risk of stroke in healthy men by 25 percent while the risk is reduced only 11 percent from other fruits. It's very interesting that consumption of vitamin C in supplement form does *not* appear to have the same benefits as the whole fruit when it comes to stroke prevention. This suggests that there must be some other protective substances in citrus juices to account for their power to protect from strokes. The current assumption at this point is that it's the power of the polyphenols that make the difference. Another reason to rely on whole foods for optimal nutrition! On the other hand, more than 350 to 400 milligrams a day of supplemental vitamin C for a period of at least ten years seems to be an effective means of lowering your risk of developing cataracts. (This is one instance where supplements do work.)

VITAMIN C SUPPLEMENTATION

I prefer whole foods for nutrients rather than relying on supplements. However, even adults consuming five servings of fruits and vegetables daily often consume less than 100 milligrams of vitamin C a day, so it's appropriate to take vitamin C supplements if you desire. Keep in mind that your body can't tell the difference between vitamin C from food and ascorbic acid made in the lab, but vitamin C in foods has polyphenols (bioflavonoids) that amplify its effect. For this reason it's best to get ascorbic acid with added bioflavonoids, giving yourself a better chance of benefiting from the whole antioxidant network. Remember also that there's a limit to how much vitamin C your body can absorb at a time, so it's best to take, say, a 250-milligram supplement in the morning and a 250-milligram supplement in the afternoon rather than a 500- or 1,000-milligram supplement at one time. If you do take supplements, be sure to keep your daily supplement intake below the Food and Nutrition Board's tolerable upper limit of 2,000 milligrams a day. In my opinion, 1,000 milligrams of supplemental vitamin C is more than enough to optimize health benefits from this vitamin.

ORANGES IN THE KITCHEN

Citrus fruit—along with cherries, grapes, and ten other fruits—won't ripen after picking. And a bright orange color doesn't necessarily mean ripe: oranges are routinely gassed and dyed for cosmetic reasons. Splotches of green on an orange are nothing to worry about.

The heavier and smaller the fruit (and, usually, the thinner the skin), the more juice it contains. You'll also get more juice out of a lemon or orange if you let it get to room temperature and roll it on the counter before juicing it.

Whole oranges can be stored either in the fridge or at room temperature. They'll last about two weeks. Don't store them in plastic bags as they can develop mold.

The amount of vitamin C in 8 ounces of orange juice can vary from about 80 to about 140 milligrams depending on the oranges and their ripeness, and on how they were processed and shipped. Heat, including pasteurization, reduces the nutrient content of juice. Check the date stamped on the carton of juice before you buy: it will stay fresh for two to four weeks once opened. Orange juice begins to lose vitamin C (and other nutrients) from the moment it's squeezed, but because the C is so abundant, as long as the juice tastes fresh, it's probably providing you with adequate amounts. One trick for revitalizing the C content of orange juice is to squeeze a lemon into the carton.

Be sure to read the labels on juices: many contain more sugar or corn syrup than juice. Only buy 100 percent fruit juice.

Here's an interesting breakdown of two sidekicks, white versus pink grapefruit, from a nutritional standpoint. It's based on one-half grapefruit:

	White grapefruit	Pink grapefruit
Calories	39	37
Vitamin C	39 mg.	47 mg.
Potassium	175 mg.	159 mg.
Lycopene	0	1.8 mg.
Beta-carotene	trace	0.7 mg.
Beta cryptozanthin	0	trace
Lutein/zeaxanthin	0	trace
Alpha-carotene	trace	trace
Flavonoids	present	present

•••

Grapefruit juice increases the bioavailability of certain drugs. It's believed that one of the flavonoids in grapefruit—likely naringin—causes this. Check with your health care provider if you are on oral medications to see if grapefruit juice will interfere.

•••

While whole fruits are the best choice, there are times when you're shopping for another form of citrus—something to spread on your toast. Citrus marmalade can be good for you: the flavonoids found in citrus fruit, which help strengthen capillaries and enhance the effects of vitamin C, survive the manufacturing process when being made into marmalade, as do many of the antioxidants and liminoids. The pectin, the soluble fiber in citrus fruit that sets marmalade, has cholesterol-lowering abilities. It's a better choice than butter on your muffin or toast!

Get some orange in your life:

- Eat an orange, tangerine, or clementine out of hand daily.
- Add mandarin orange segments to a spinach salad with some chopped red onion.
- Sprinkle grapefruit halves with a dusting of brown sugar and broil for a great dessert.
- Add orange juice to a fruit smoothie.
- Keep some orange and/or lemon zest in your freezer: put it into cakes, cookies, muffins, or even drinks for a refreshing boost of nutrition and flavor. Sprinkle it on yogurt, into fruit salads and even chicken salad. Use citrus zest in hot tea. Citrus juice adds a flavor boost to many poultry and fish dishes.
- Bring back orange wedges as a refresher for athletes, young and old. They give a much-needed boost of antioxidants and vitamin C on the playing field or at the gym.

••

Kumquats are the smallest of the citrus fruits. I'm fortunate enough to have two kumquat bushes outside my front door and when they're in season, I pop one or two a day into my mouth. The thin, sweet rind is jammed with powerful phytonutrients. You can find them in supermarkets, mostly in the winter. Pinch the fruit between your fingers before eating; you'll release the juice and get a blast of sweet and tart from this little nutrition powerhouse.

••

ORANGE BRAN FLAX MUFFINS

YIELDS 24 MUFFINS

1 ½ cups oat bran
1 cup all-purpose flour
1 cup ground flaxseed
1 cup natural bran
1 tablespoon baking powder
½ teaspoon salt
2 oranges, washed, quartered, and seeded
¾ cup brown sugar
1 cup buttermilk
½ cup canola oil
2 eggs
1 teaspoon baking soda
1 ½ cups raisins

In a large bowl, combine the oat bran, flour, flaxseed, bran, baking powder, and salt. Set aside. In a blender or food processor, combine the oranges, brown sugar, buttermilk, oil, eggs, and baking soda. Blend well. Pour the orange mixture into the dry ingredients. Mix until well blended. Stir in the raisins (white chocolate chips can be substituted for the raisins). Fill paper-lined muffin tins almost to the top. Bake in a 375°F oven for 18 to 20 minutes, or until a wooden pick inserted in the center of the muffin comes out clean. Cool in the tins for 5 minutes before removing to a cooling rack.

•••

Discovered by a Canadian housewife who borrowed one of her husband's woodworking tools and was delighted with the results when she used it to zest a lemon, the Microplane zester makes it very easy to zest any citrus. You can save the zest in plastic bags in the freezer to use in a variety of recipes. The zesters are sold in many kitchen stores and online.

•••

FRESH ORANGE LEMONADE

MAKES TWELVE 8-OUNCE SERVINGS

2½ cups water
1 cup sugar (or less)
2 tablespoons orange zest
2 tablespoons lemon zest
1½ cups fresh orange juice, about 5 oranges
1½ cups lemon juice, about 8 lemons
Water
Ice cubes
Citrus slices, optional

In a medium saucepan, combine the water and sugar. Cook over medium heat until the sugar dissolves, stirring occasionally. Remove from the heat and cool. Add the orange and lemon zests and juices to the sugar mixture. Cover and let stand at room temperature for 1 hour. Cover and refrigerate until serving time.

To serve, fill glasses with equal parts fruit mixture and water. Add ice and serve. If you like, garnish the lemonade with citrus slices.

Pumpkin

SIDEKICKS: carrots, butternut squash, sweet pota-toes, orange bell peppers
TRY TO EAT: ½ cup most days

••

Pumpkin contains:

- Alpha-carotene
- Beta-carotene
- High fiber
- Low calories

- Vitamins C and E
- Potassium
- Magnesium
- Pantothenic acid

••

"Pumpkin?" people ask me. "Why pumpkin?" Most people find pumpkin the dark horse of the SuperFoods. Many of us rarely think of pumpkin as a food. We buy a pumpkin to carve at Halloween when it serves as a glorified candleholder that's disposed of once the trick or treaters go home. We only eat it once a year, if at all, in a Thanksgiving pie. Most people think of pumpkin as a decorative gourd rather than a highly nutritious and desirable food.

This is unfortunate because the squash known as pumpkin is one of the most nutritionally valuable foods known to man. (By the way, pumpkin is not a vegetable; it's a fruit. Like melons, it's a member of the gourd family.) Moreover, it's inexpensive, available year round in canned form, incredibly easy to incorporate into recipes, high in fiber, and low in calories. All in all, pumpkin is a real nutrition superstar.

Native Americans cherished the pumpkin and used both the flesh and the seeds as a dietary staple—the former as a reliable food, fresh roasted or dried, and the latter as medicine. By the second Thanksgiving, pumpkin had become one of the main attractions of the annual feast. In fact, the early settlers made a version of pumpkin pudding that resembles one of my favorite recipes, which I'll include in this chapter. They took a pumpkin, filled it with milk, spices, and honey, and baked it in hot ashes. The resulting sweet dish resembles Patty's Pumpkin Pudding (page 128), although I doubt it was as delicious.

The nutrients in pumpkin are really world class. Extremely high in fiber and low in calories, pumpkin packs an abundance of disease-fighting nutrients, including potassium, pantothenic acid, magnesium, and vitamins C and E. The key nutrient that boosts pumpkin to the top of the SuperFoods Rx list is the synergistic combination of carotenoids. Pumpkin contains one of the richest supplies of bioavailable carotenoids known to man. Indeed, a half-cup serving of pumpkin gives you *more than two times* my recommended daily dietary intake of alpha-carotene and *100 percent of* my recommended daily dietary goal of beta-carotene. When you realize the tremendous benefits of these nutrients, you'll see why pumpkin is such an extraordinary nutrition superstar.

Daily Carotenoid Recommendation

The Food and Nutrition Board of the Institute of Medicine of the National Academy of Sciences is charged with setting the recommended daily allowances for various nutrients. While they have recognized that "higher blood concentrations of beta-carotene and other carotenoids obtained from foods are associated with lower risk of several chronic diseases," as yet they have been unable to arrive at a recommended daily intake of carotenoids. In the meantime, I've come up with recommendations, based on all the available peer-reviewed literature, that I feel ensures that you are consuming the optimum daily protective amounts of these nutrients.

Alpha-carotene: 2.4 milligrams or more from food sources
Beta-carotene: 6 milligrams or more from food sources
Lucopene: 22 milligrams or more from food sources
Lutein zeaxanthin: 12 or more from food sources
Beta cryptoxanthin: 1 milligram or more from food sources

• •

Carotenoids are deep orange-, yellow-, or red-colored, fat-soluble compounds that occur in a variety of plants. They protect the plants from sun damage while they help them attract birds and insects for pollination. So far scientists have identified about six hundred carotenoids, and more than fifty of them commonly occur in our diet. Not all dietary carotenoids are efficiently absorbed. As a result, only thirty-four carotenoids have currently been found in our blood and human breast milk. The six most common carotenoids found in human tissue include beta-carotene, lycopene, lutein, zeaxanthin, alpha-carotene, and beta cryptoxanthin. Both alpha- and beta-carotene and beta cryptoxanthin are what's known as pro–vitamin A carotenoids, which means that the body can convert them to vitamin A. As opposed to animal sources of vitamin A, these plant sources cannot deliver a toxic amount of the vitamin. (Interestingly, if you eat enough polar bear liver, which is extremely rich in so-called preformed animal vitamin A, you can die from a toxic dose of that vitamin.) Carotenoids are concentrated in a wide variety of tissues, where they help protect us from free radicals, modulate our immune response, enhance cell-to-cell communication, and possibly stimulate production of naturally occurring detoxification enzymes. Carotenoids also play a major role in protecting the skin and eyes from the damaging effects of ultraviolet light.

Foods rich in carotenoids have been linked to a host of health-promoting and disease-fighting activities. They have been shown to decrease the risk of various cancers, including those of the lung, colon, bladder, cervical, breast, and skin. In the landmark Nurses' Health Study, women with the highest concentrations of carotenes in their diets had the lowest risk of breast cancer.

••

Quick Carotenoid Boosts

To get your daily dose of carotenoids, enjoy a sliced orange or red bell pepper, peeled baby carrots—which can be quickly steamed or microwaved or eaten raw—a handful of fresh or dried apricots or prunes, a slice of cantaloupe or watermelon, a slice of mango, or a persimmon. One other carotenoid boost that everyone loves: Häagen-Dazs mango sorbet.

••

Carotenoids have also shown great promise in their ability to lower rates of heart disease. In one thirteen-year-long study, researchers found a strong correlation between lower carotenoid concentrations in the blood and a higher rate of heart disease. As has frequently been found, the correlation between increased carotenoid consumption and decreased risk of heart disease was higher when all carotenoids, not just beta-carotene, were considered.

Carotenoid consumption also decreases the risk of cataracts and macular degeneration.

The two carotenoids that are richly present in pumpkin—beta- and alpha-carotene—are particularly potent phytonutrients.

Beta-carotene, which first came to attention in the 1980s, is one of the world's most studied antioxidants. The word "carotenoid"—derived from "carrot"—comes from the yellow-orange color of these nutrients, which at first were linked primarily with carrots. Carrots (and sweet potatoes) also contain rich amounts of beta-carotene. It's abundant in fruits and vegetables, and we've long known that the beta-carotene in foods helps prevent many diseases, including lung cancer. It was the connection between beta-carotene and lung-cancer prevention that led to some fascinating studies. These groundbreaking studies on beta-carotene were among the first indications that supplements were not the total answer in preventing disease and, indeed, it's this finding that's at the heart of SuperFoods: whole foods are the answer to disease prevention and health promotion.

Scientists reasoned that if the beta-carotene in foods helped to prevent lung cancer, it followed that a beta-carotene supplement would do the same. Unfortunately, and shockingly, two important studies showed that, to the contrary, smokers who took beta-carotene supplements showed an increase in lung cancer.

Perhaps you recall those studies. They made front-page news around the world a few years ago:

- In 1996, a Finnish study on 29,000 male smokers, published in the *New England Journal of Medicine*, showed that those who smoked and took beta-carotene supplements were 18 percent *more* likely to develop lung cancer than those who had not taken supplements.

- In the United States, the Carotene and Retinal Efficacy Trial (CARET) study, which was published in the *Journal of the National Cancer Institute*, was halted almost two years before expected completion because of the negative effects of the supplemental beta-carotene and vitamin A on smokers when compared with subjects taking a placebo.

The news of the surprising outcomes was stunning to those who follow health trends. They had become accustomed to regular positive reports that individual micronutrients help to prevent disease. What went wrong? In simplest terms, the beta-carotene found in foods, working synergistically with the other nutrients present in that food, have a very different effect on the body than a single nutrient isolated from its web of assisting and augmenting synergistic partners. The carotenoids, like many nutrients, work best as a team; break up the team and results can be unpredictable.

Moreover, the dose of a nutrient in supplement form differs from the dose you'd get from food. For example, we know that 10 milligrams of beta-carotene daily from carrots is good for you and helps prevent disease. But if you give someone 20 milligrams of beta-carotene in a pill, the dose may behave more like a drug than a nutrient, with unintended and possibly adverse health consequences. Why? Because the absorption rate of beta-carotene in uncooked carrots is only about 10 percent. If you cook the carrot, the absorption rate of the beta-carotene in it goes up to about 29 percent. So if you're getting your beta-carotene from carrots, your body is only absorbing a percentage of it. You can't eat enough carrots to get a toxic dose. There is no toxic dose of carotenoids from carrots, pumpkin, or other foods. In fact, the only known (harmless) side effect of ingested whole food carotenoids is an orange (beta-carotene) or reddish (lycopene) tinge to the skin. On the other hand, if you take beta-carotene in supplemental form, your body is absorbing a very high percentage of this micronutrient. Not only is it suddenly a potentially pharmacological dose of a nutrient, it's a dose that may throw off-balance the synergy of the other nutrients your body depends on to maintain health.

When derived from whole foods like pumpkin, the carotenoids are major players in the fight against disease. Higher blood levels of beta-carotene and alpha-carotene are associated with lower levels of certain chronic diseases. In laboratory studies, beta-carotene has been shown to have very powerful antioxidant and anti-inflammatory properties. It prevents the oxidation of cholesterol in laboratory studies and, since oxidized cholesterol is the type that builds up in blood vessel walls and contributes to the risk of heart attack and stroke, getting extra beta-carotene in the diet may help to prevent the progression of atherosclerosis and heart disease.

Beta-carotene along with other carotenoids may also prove to be helpful in preventing the free radical–caused complications of long-term diabetes and the increased risk for cardiovascular disease associated with this common illness.

Studies have also shown that a good intake of beta-carotene can help to reduce the risk of colon cancer, possibly by protecting colon cells from the damaging effects of cancer-causing chemicals.

While beta-carotene has long been linked with health promotion, it's the bounty of alpha-carotene in pumpkin that makes it a real nutrition standout. The exciting news about alpha-carotene is that its presence in the body along with other key nutrients is reportedly inversely related to biological aging. In other words, the more alpha-carotene you eat, the slower your body shows signs of age. Not only may alpha-carotene slow down the aging process, it also has been shown to protect against various cancers and cataracts. Moreover, the combination of carotenoids, potassium, magnesium, and folate found in pumpkin offers protection against cardiovascular disease.

Pumpkin is also a terrific source of fiber. Most people aren't aware of the fiber content of canned pumpkin because it seems so creamy. Just one half-cup serving provides 5 grams of fiber—more than you're getting from most supermarket cereals.

••

Alpha-Carotene All-Stars

Pumpkin (cooked, 1 cup)	11.7 mg.
Carrots (cooked, 1 cup)	6.6 mg.
Butternut squash (cooked, 1 cup)	2.3 mg.
Orange bell pepper (1 cup)	0.3 mg.
Collards (cooked, 1 cup)	0.2 mg.

Beta-Carotene All-Stars

Sweet potato (cooked, 1 cup)	23 mg.
Pumpkin (cooked, 1 cup)	17 mg.
Carrots (cooked, 1 cup)	13 mg.
Spinach (cooked, 1 cup)	11.3 mg.
Butternut squash (cooked, 1 cup)	9.4 mg.

••

PUTTING PUMPKIN IN YOUR DIET

It's all very well that pumpkin is such a nutritional power-house, but that doesn't count for much if every time you wanted some pumpkin you had to wrestle one of those big orange gourds into the kitchen. A winter squash, pumpkin is usually available fresh only in the autumn and early winter and the rest of the year you might have trouble finding one. But one of the best features of pumpkin is that it's readily available all year long in an inexpensive canned form. At our house, my wife Patty's pumpkin pudding is always on hand. Our kids all love it and their visiting friends, who are sometimes skeptical of "healthy" foods, dig right in when we serve it.

Canned pumpkin is one of those foods that give the lie to the notion that fresh is always best. Not only is it some-times difficult if not impossible to find fresh pumpkin, canned pumpkin is actually *more* nutritious (except for those pumpkin seeds: see page 127). Canned pumpkin puree (don't get it mixed up with "pumpkin pie filling," which has added sugar and spices) has been cooked down to reduce the water content that you'd find in fresh pumpkin. At only 83 calories a cup, it offers more than 400 percent of my recommendation of alpha-carotene and close to 300 percent of my beta-carotene recommendation, as well as almost half of the iron requirement for adult men and postmenopausal women.

OTHER WINTER SQUASH

Pumpkin is not the only winter squash that's filled with beneficial carotenes. There are a number of winter squashes that are available in the market for most of the year that come close to pumpkin in their nutritional benefits. Don't make the mistake of thinking that all hard winter squashes taste alike. Not so. Try to experiment with different varieties. Most people are familiar with acorn squash, but it is often tasteless. Try butternut squash (which is highly nutritious and makes a great soup), buttercup squash (it looks like it's wearing a "hat" where the stem is), delicata squash (looks like a fat yellow-orange cucumber with green stripes), or hubbard squash (deep green, more rounded than an acorn squash).

There's a vast difference between an over-the-hill watery acorn squash and a fresh, delicious ripe one. Buy these squashes at farmers' markets where you know they'll be fresh since they come from local producers. In any case, here are some tips for finding the best-tasting and most nutritious winter squashes:

- A winter squash should be rock hard. If it's soft, it's either too young or too old. Test the skin: if it nicks easily, it's probably too young.
- Look for squash with the stem on. Without the stem, bacteria can get into the squash.
- The skin should be somewhat dull. A shiny squash is either too young or has been treated with wax.
- A deep, rich color usually means a ripe squash. If the squash is dark green, you can still usually see the area that touches the ground and it should be a ripe color, not pale green.
- You'll find the most vivid colors at harvest time—usually late summer to fall. But later in the year, when the squash has been stored, it will be sweeter and more concentrated in flavor.
- Squash that comes from a cooler climate will often have more flavor and sweetness than one that grows in a warmer place. Check your supplier.

Mix canned pumpkin with low-fat or nonfat yogurt or with applesauce. You can drizzle it with buckwheat honey and a few raisins. Use canned pumpkin in recipes for soups, breads, and muffins.

PUMPKIN SIDEKICKS

While pumpkin is a flagship SuperFood, there are other terrific choices in this category that provide a bounty of carotenoids as well as other nutritional benefits. Carrots are perhaps the most popular. The nutrients in carrots are more bioavailable when cooked; so, while there's nothing wrong with eating raw carrots, you will get more nutritional benefit from them when they're cooked.

Pumpkin Seeds

You can buy pumpkin seeds—often labeled pepitas, "little seeds" in Spanish—or you can scoop the seeds out of a fresh pumpkin and toast them. They're rich in vitamin E, iron, magnesium, potassium, and zinc. Moreover, they are a great plant-based source of omega-6 and omega-3 fatty acids. Remove any pulp or strings from the pumpkin seeds and rinse them in fresh water. Air-dry them by letting them sit overnight and drizzle them with a bit of olive oil and a sprinkle of sea salt. Roast on a cookie sheet in a 350°F oven for 15 to 20 minutes. Sprinkle with curry powder or chili powder, if you like. Cool completely and store in an airtight container in the fridge.

Sweet potatoes are another excellent source of beneficial carotenoids. Pierce them all over with a fork and pop them in the microwave for about five minutes if you don't have time to oven-bake them. If oven-baked they begin to caramelize as they cook. A baked sweet potato doesn't even need butter—just a bit of salt and pepper. I often brown-bag half a baked sweet potato to eat cold at work. It tastes great at room temperature if you don't have a microwave in the office to heat it up. Another sweet potato treat is to slice them thin (a mandolin helps here), toss them with a bit of olive oil and some coarse salt, and roast them on a cookie sheet in a 400°F oven for about 20 minutes, turning them a few times. Be careful not to burn them and keep in mind that thinner slices will cook more quickly.

PATTY'S PUMPKIN PUDDING

½ cup sugar
1 teaspoon cinnamon
½ teaspoon salt
¼ teaspoon ground ginger, optional
¼ teaspoon ground cloves, optional
2 large eggs (with high omega-3 content, as noted on label)
One 15-ounce can Libby's 100 percent pure pumpkin
One 12-ounce can evaporated nonfat milk (or evaporated 2 percent milk)

Mix sugar, cinnamon, salt, ginger, and cloves in a small bowl. Beat the eggs in a large bowl. Stir in the pumpkin and sugar-spice mixture. Gradually stir in the evaporated milk. Pour into a shallow ovenproof dish and bake in a preheated 350°F oven for about 40 minutes. Don't overbake; the center should be slightly wiggly. Cool and enjoy at room temperature or refrigerate for later use.

ROASTED BUTTERNUT SQUASH WITH BUCKWHEAT HONEY GLAZE

MAKES 2 SERVINGS

This is great with a turkey or salmon dinner.

Canola oil
1 medium butternut squash, cut in half lengthwise,
    strings and seeds scooped out and discarded
1 tablespoon butter
2 tablespoons buckwheat honey
Salt and freshly ground black pepper

Preheat the oven to 400°F. Lightly grease a baking sheet with canola oil. Put the squash halves, cut side down to help them caramelize, on the sheet. Bake for 45 to 55 minutes, or until a skewer easily pierces the squash. In the meantime, melt the butter and add the honey with some pepper and salt to taste. When the squash is cooked, remove it from the oven and turn the halves cut side up. Brush them with the honey-butter mixture and return them to the oven for about 5 minutes. Cut the squash into serving pieces and serve.

••

Four tablespoons of buckwheat honey contain about 10 milligrams of polyphenols. That may not seem like a lot, but a study of twenty-five healthy males found a 7 percent increase in their serum antioxidant capacity after drinking warm water with 4 tablespoons of buckwheat honey dissolved in it.

••

Wild Salmon

SIDEKICKS: Alaskan halibut, canned albacore tuna, sardines, herring, trout, sea bass, oysters, and clams
TRY TO EAT: fish two to four times per week

••

Salmon contains:

- Marine-derived omega-3 fatty acid
- B vitamins
- Selenium
- Vitamin D
- Potassium
- Protein

••

Once upon a time (actually not very long ago), people came to believe that fat was a murderous monster and the ideal diet was completely devoid of any fat whatsoever. It was the era of fat-free. Fat-free salad dressings, nonfat cakes and cookies, no-fat soups and casseroles. Even bottles of fruit juices proudly trumpeted "a fat-free food" on their labels. (Was there ever a fatty cranberry juice?) Why this fear of fat? It all started as a well-intentioned campaign to improve health. The second half of the twentieth century saw an alarming epidemic of heart disease. Countless studies sought reasons for this epidemic. It became clear that smoking, a sedentary lifestyle, and a high-fat diet were linked to the rising tide of cardiovascular disease. The lesson was obvious: to reduce your risk of heart disease, a major killer, you should cut as much fat as possible out of your diet. Cholesterol became a household word and Americans became fat-phobic.

It's taken years for the more complicated and interesting truth to emerge. First, research indicating that all fat is not bad began to reach the public. We all needed an education in dietary fat and, bit by bit, we got one. In a nutshell, we learned that we derive four basic types of fat from food: saturated fat, trans fat (partially hydrogenated oils), monounsaturated fat, and polyunsaturated fat. The news on saturated fat hasn't changed: saturated fat—found primarily in red meat, full-fat dairy products, and some tropical oils—has well-established negative health effects, increasing your risk of diabetes, coronary heart disease, stroke, some cancers, and obesity. One researcher, writing in the *Journal of the American Dietetic Association*, concluded that "reducing dietary intake of saturated fatty acids may prevent thousands of cases of coronary heart disease and save billions of dollars in related costs." There's little positive about saturated fat and it should make up no more than 7 percent of your fat calories per day.

Trans fats—listed on food labels as "partially hydrogenated vegetable oil"—are also bad, probably even worse than saturated fat. Trans fats were created by chemists seeking a fat that would store better than animal fats. They were an attempt to lengthen the shelf life of foodstuffs.

Remember, there are good fats. The good guys in the fat family are the monounsaturated fats—the kinds found in olive and canola oils. These fats not only protect your cardiovascular system, they also lower the risk of insulin resistance, a physiologic state that can lead to diabetes and possibly cancer.

••

The average intake of trans-fatty acids from partially hydrogenated vegetable oils is currently about 3 percent of total calorie intake per day. Currently there is no recommended safe intake of these oils. The Nurses' Health Study suggests that the incidence of type II diabetes could be reduced by 40 percent or more if these fats were consumed in their original, unhydrogenated form.

••

Finally, we come to polyunsaturated fatty acids. Both omega-6 (linoleic, or LA, fat) and omega-3 (alpha linolenic, or ALA, fat) are so-called essential polyunsaturated fatty acids (EFAs). Our bodies cannot manufacture these two fats and therefore we must rely on dietary intake to avoid a deficiency in these essential (for life) fats. Omega-6 fatty acids are currently overabundant in the typical Western diet. They are present in corn, safflower, cottonseed, and sunflower oils. Virtually no one in America is deficient in these ubiquitous fatty acids. If you look at almost any packaged food, you're going to see one of these oils as an ingredient.

Let's look for a minute at the omega-3 class of polyunsaturated fat. Omega 3 fats come in two distinct forms: plant derived (ALA) and largely marine species derived (EPA/DHA). With each passing month, additional studies are being published about the health benefits of omega-3s. Unfortunately, many Americans are currently deficient in the omega-3 class of essential fatty acids. Omega-3 fatty acids—the ones that help make salmon a SuperFood—haven't been included in adequate amounts in our diet, partly because of lack of knowledge on the part of the public and also because they've been "processed out" of our modern diet. This deficiency has long-term and disastrous health consequences for many people. Indeed, William S. Harris, writing in the *American Journal of Clinical Nutrition*, has said: "In terms of its potential impact on health in the Western world, the Omega 3 story may someday be viewed as one of the most important in the history of modern nutritional science." Dr. Evan Cameron, from the Linus Pauling Institute, has said: "Our epidemic of heart disease and cancer may be the result of a fish oil deficiency so enormous we fail to recognize it." The bottom line: it's not just okay to include omega-3 fatty acids in your diet, it's imperative to do so if you want to restore a critical balance in your body that is most likely out of whack.

Enter salmon. Salmon is one of the richest, tastiest, readily available sources of marine-derived omega-3 fatty acids. By including wild salmon (or its sidekicks) in your diet two to four times a week (see "Tuna Guidelines," page 137) you should achieve optimal protection against a multitude of diseases that have been associated with low intakes of these critical fats.

THE CRITICAL BALANCE OF EFAS

The key to EFAs—as with so many health issues—is balance. Your body can't function optimally without a balanced ratio of EFAs. The optimum balance of essential fatty acids is a balance of omega-6 to omega-3 that is somewhere between 1 to 1 and 4 to 1. Unfortunately, the typical Western diet contains fourteen to twenty-five times more omega-6 than omega-3 fatty acids. This unbalanced ratio that most of us live with determines myriad biochemical events that affect our health. For example, too much omega-6 (the oil that dominates our typical diet) promotes an inflammatory state, which in turn increases your risk for blood clots and narrowing of blood vessels.

• •

To get a healthy amount of omega-3–omega-6 EFAs in your diet:

- Use omega-3 enriched eggs.
- Cook with canola rather than corn or safflower oil.
- Eat soy nuts and walnuts.
- Sprinkle wheat germ on cereal and yogurt; use it in baking.
- Eat wild salmon or its sidekicks two to four times per week.
- Look for salad dressings with at least some soybean or canola oil.
- Use flaxseed oil (stored in a dark bottle; refrigerated constantly; discarded after a couple of months) sparingly in homemade salad dressings.
- Use ground flaxseed in muffins, breads, and pancakes.
- Avoid processed and refined foods whenever possible, including packaged cakes, cookies, and baked goods.

• •

We now also know that without sufficient intake of omega-3 fatty acids, the body cannot adequately build an ideal cell membrane. Membranes that are poorly constructed are not capable of optimizing cellular health, which in turn increases your risk for a host of health problems, including stroke, heart attack, cardiac arrhythmias, some forms of cancer, insulin resistance—which can lead to diabetes—asthma, hypertension, age-related macular degeneration, chronic obstructive lung disease (COPE), autoimmune disorders, attention deficit hyperactivity disorder, and depression. This list seems to encompass the major ailments of the twentieth century. Some scientists have claimed that the proliferation of these disorders is at least in part due to the lack of omega-3 fatty acids in our diet.

One report estimated that close to 99 percent of Americans do not consume enough omega-3 fatty acids, and 20 percent of us have such low levels of omega-3 fatty acids that they can't even be detected. This EFA deficiency is rarely noticed because the symptoms it produces are so vague. Dry skin, fatigue, brittle nails and hair, constipation, frequent colds, inability to concentrate, depression, and joint pain can all result from lack of omega-3 fatty acids in the diet. Many of us live with these conditions, never dreaming that we may be suffering from a nutritional deficiency that can ultimately cause serious chronic disease and even death.

••

What About Cod Liver Oil?

Patients have asked me if a tablespoon of cod liver oil is the solution to low levels of omega-3 fatty acids. Yes and no. While it's true that cod liver oil contains a good dose of omega-3 fatty acids (which probably protected our grandparents from many ailments), it's also true that, in addition to tasting pretty awful, it may be contaminated by mercury and PCBs.

••

OUR DETERIORATING DIET

The decline of omega-3 fatty acids in our diet is an interesting story. Up until the twentieth century, this group of fatty acids was abundant in our foods. Some scientists even postulate that it was the ingestion of omega-3 EFAs that allowed the brain to evolve to the next stage in human development. Not only found in cold-water fish, omega-3 fatty acids were available in green leafy foods (we eat one-third of the greens eaten by our ancestors) as well as meat from animals that fed on grass (rather than the grain-fed meat we now consume which is low in omega-3 fatty acids). As foods became more processed, the amounts of omega-3 fatty acids diminished while the amounts of omega-6 fatty acids increased to today's dangerous ratios. In fact, seventy years ago, before solvent-extracted vegetable oils and corn-based animal husbandry, people were not exposed to the high intake of omega-6 fatty acids we experience today. As one researcher said, "We may well be experiencing the 'linoleic acid paradox' in which a supposedly healthy fatty acid (i.e., one that lowers total cholesterol) is associated with increasing rates of cancer, inflammatory and cardiovascular diseases during those same decades. Compounding and confounding this paradox are low intakes of ALA (plant-based omega-3 oils) and other omega-3 fish oils."

People who eat diets with the optimum balance of essential fatty acids manage to avoid many common ailments. Eskimos in Greenland first brought attention to the question of fat in the diet because they had little heart disease despite a diet high in fat (40 percent of their total calorie intake, which included more than 10 grams of EPA/DHA daily). The Lyon Heart Study compared the effects of a modified Crete diet—one enriched with omega-3 fatty acids—and the American Heart Association diet. This landmark study showed a reduction of risk, 56 percent for total deaths and 61 percent for cancers, in the experimental group (which ate a diet high in ALA) compared to the control group. The Japanese, who rely on a traditional diet of fish, are likewise protected from heart disease while their neighbors who farm and eat far less fish suffer higher levels of cardiovascular disease. It's interesting to note that cultures that have high omega-3 consumption in fish have far less depression than those whose diet is dominated by omega-6 fatty acids. In fact, in one fascinating epidemiological study, fish consumption was the most significant variable in comparing levels of depression and coronary heart disease.

The bottom line: there is a critical and optimal balance of omega-3 and omega-6 essential fatty acids, working with the nutritional cofactors of minerals, vitamins, phytonutrients, fiber, antioxidants, and electrolytes, which is important in reducing the incidence of many of the degenerative diseases currently epidemic in Western countries.

••

Tuna Guidelines

Canned tuna is a popular source of omega-3 EFAs.

Some tips on using canned tuna in your diet:
- Because of the potential mercury content adults shouldn't eat more than one can of tuna a week.
- Buy albacore tuna—it's the richest tuna source of omega-3 EFAs.
- Buy tuna packed in spring water so you won't be getting extra fat.
- Low-salt canned tuna is best.

••

BENEFITS OF OMEGA-3 FATTY ACIDS

Here are some of the benefits you can enjoy from increasing your omega-3 fatty-acid intake by following the Super-Foods recommendation and adding wild salmon and other cold-water fish to your diet:

- Reduce your risk of coronary artery disease. We know that omega-3 fatty acids help to increase your HDL ("good" cholesterol), reduce your blood pressure, and stabilize your heartbeat, thus preventing one of the causes of fatal heart attacks: sudden cardiac arrhythmias. Omega-3 EFAs also act as blood thinners, reducing the "stickiness" of platelets that can lead to clots and stroke. In one study, patients who had a heart attack and who were given 1 gram per day of omega-3s had a 20 percent drop in total mortality, a 30 percent drop in death from cardiovascular disease, and a 45 percent drop in sudden death compared with similar patients who were given nothing or vitamin E alone.

- Control hypertension. The bottom line is the more omega-3 fatty acids you eat, the lower your blood pressure. This is because of the beneficial effect of omega-3s on your artery walls' elasticity. A 1993 meta-analysis on the effects of fish oil on blood pressure showed that eating cold-water fish three times a week seems to be as effective as high-dose fish oil supplements in reducing blood pressure in hypertensive patients.
- Prevent cancer. Research is just beginning to demonstrate that omega-3 fatty acids may play a role in preventing both breast and colon cancers.
- Prevents age-related macular degeneration. In the Nurses' Health Study, those who ate fish four or more times a week had a lower risk of age-related macular degeneration than those who ate three or fewer fish meals per month. The most prevalent fatty acid in our retina is DHA, and the primary dietary source of this "good fat" is salmon and other so-called heart-healthy fish. DHA also seems to reduce some of the adverse effects of sunlight on retinal cells.
- Mitigate autoimmune diseases such as lupus, rheumatoid arthritis, and Raynaud's disease. Researchers believe that the anti-inflammatory abilities of omega-3 fatty acids are what help reduce the symptoms of autoimmune diseases as well as prolong the survival of those who

suffer from them. Multiple studies have substantiated these results.

- Relieve depression and a host of mental health problems. Perhaps the most interesting research on omega-3 fatty acids involved their relationship to mental health ailments such as depression, attention deficit hyperactivity disorder, dementia, schizophrenia, bipolar disorder, and Alzheimer's disease. Our brains are surprisingly fatty: over 60 percent of the brain is fat. Omega-3 fatty acids promote the brain's ability to regulate mood-related signals. They are a crucial constituent of brain-cell membranes and are needed for normal nervous system function, mood regulation, and attention and memory functions.

HOW MUCH OMEGA-3 SHOULD YOU EAT?

There are some overlapping and some individual properties unique to each type of omega-3 fatty acid (ALA and EPA/DHA). Therefore, it's better to have a combination of both in your diet. There are no current published human clinical trials telling us the ideal intake ratio of ALA versus EPA/DHA. Until such studies are available, I recommend a combination of both.

While we often think that if a little of something is good, a lot must be better, be cautious: too much omega-3 fatty acid can promote a risk of stroke by thinning the blood excessively. Bleeding time is prolonged with an intake of omega-3 fatty acids that exceeds 3 grams a day. (Greenland Eskimos who consume an average 10.5 grams a day of omega-3 fatty acids have an increased risk of hemorrhagic stroke.) Too high a daily dose can also negatively affect your immune system. However, a study published in the May 2003 *American Journal of Clinical Nutrition* found that an intake of 9.5 grams or less of ALA or 1.7 grams or less of EPA/DHA did not alter the functional activity of three important cells involved in inflammation and immunity. People on blood thinners and/or aspirin should take this into consideration when they alter their fatty acid intake and should consult their health care professional.

The Food and Nutrition Board of the Institute of Medicine, the National Academies, recently revised the recommended daily intake of ALA (plant-derived omega-3) to 1.6 grams for adult men and 1.1 grams for adult women. They didn't feel it was possible to set an acceptable range for all omega-3 fatty acids (ALA, DHA, EPA). They therefore recommended that a target amount of EPA or DHA is 160 milligrams a day for men and 110 milligrams a day for women. I personally feel the EPA/DHA recommendation should be higher to achieve optimal health. My goal is about 1 gram of marine-derived EPA/DHA a day and I concur with the Food and Nutrition Board's daily recommendation on ALA, which is the amount found in less than 1 tablespoon of flaxseed.

WILD SALMON TO THE RESCUE

Some of my patients roll their eyes when I begin to talk about the balance of omega-6 EFAs and omega-3 EFAs. Frankly, they're not very interested in the biochemistry of fat; they just want to know easy ways to improve their health. When it comes to omega-3 fatty acids, wild salmon is one simple answer. Add it to your diet. Wild salmon is delicious, high in protein, widely available in canned form, easy to prepare, and, more important, high in beneficial omega-3 fatty acids. If you eat wild salmon or other cold-water fish, like sardines or trout, two to four times a week, and incorporate some of the other recommendations in this section about the use of oils, you will "rebalance" the ratio of fatty acids in your body and be on your way to vastly improving your cellular health. There's ample evidence that including cold-water fish like wild salmon in your diet will have a positive effect on your short- and long-term health. Keep in mind, it can take up to four months to achieve an ideal omega-3 fatty acid concentration in your body. The American Heart Association currently recommends eating two servings of fish a week (preferably fatty fish like wild salmon) for people without a history of coronary heart disease and more if you do have such a history. I think three to four servings a week gives a broad range of protection against a wide variety of chronic diseases, but eat a variety of these fishes, too.

THE "D" DILEMMA

Vitamin D deficiency is a major, largely unrecognized epidemic in adult men and women in the United States today. In one survey of healthy people in Boston between the ages of eighteen and twenty-nine, 36 percent were found to be vitamin D deficient. Another study reported that 42 percent of African-American women ages 15 to 49 and 4.2 percent of white females in the same age group are deficient in vitamin D. This is important because there seems to be an inverse risk of dying from breast, colon, ovarian, and prostate cancers with a decreased presence of vitamin D. African Americans, many of whom are chronically vitamin D deficient, have a higher incidence and more aggressive forms of many cancers, including breast and prostate cancers. (It's important to recognize that, while we call it a vitamin, vitamin D really acts more like a hormone in our bodies.) Studies indicate that men who are exposed to sunlight can delay the onset of prostate cancer by more than five years, and children receiving vitamin D supplementation beginning at the age of one year had an 80 percent decreased risk of developing type I diabetes. Adequate vitamin D intake is associated with a lower risk of hip fractures due to osteoporosis in postmenopausal women. In one study on this problem, neither milk nor a diet high in calcium seemed to reduce the risk.

The most important source of vitamin D is the skin's synthesis of the vitamin from sunlight exposure. People who live far from the equator (and therefore get less sunlight), who use sunblocks, or who have heavy skin pigmentation (African Americans have high melanin concentrations, which limit vitamin D synthesis), all may be at risk for low levels of vitamin D. Sunblocks can cut vitamin D production by about 95 percent. I am not recommending discontinuing using sunscreen, but this is a good reason to increase your dietary sources of vitamin D. Interestingly, vitamin D deficiency did not become a health problem until after the onset of the industrial revolution, which led to a decrease in exposure to sunlight as more and more people worked indoors. The major food sources of vitamin D include fatty fish like sardines, salmon, and tuna, and fortified foods, particularly cereals and some dairy products.

How do you protect yourself from vitamin D deficiency? Add wild salmon, sardines, and tuna to your diet. Try to get about fifteen minutes of sun exposure to your arms and face at least three times a week, before 10 A.M. and after 3 P.M., when the ultraviolet rays are not as damaging to the skin. Also check the labels of any fortified foods you eat, including cereals, milk, and soymilk. The Food and Nutrition Board has been unable to settle on an RDA for vitamin D. The current recommendations for adequate intakes are: adult males and females ages 19 to 50 should get 200 IU per day; ages 51 to 70 should get 400 IU a day; over 70 years, 600 IU per day. If you don't get enough vitamin D from all your food sources, you might consider taking a supplement, though you should be careful with vitamin D supplementation, as there is a definite risk of toxicity if you take too much.

A FISH STORY

Wild salmon, and all fish for that matter, used to carry a reliable nutritional benefit. The fish, in their natural habitat, love to eat zooplankton (tiny single-celled organisms), which are a rich source of omega-3 fatty acids. People who ate the fish thus delivered this healthy fat to their eagerly awaiting cells. Sadly, as the oceans have become overfished and polluted, the picture has changed. For one thing, U.S. Atlantic salmon are virtually extinct. (Most Atlantic salmon sold in the United States is farm raised.) Even worse from a health standpoint, some cold-water fish are contaminated with mercury. These include swordfish, shark, tilefish, and king mackerel. Avoid eating these fish.

Today, farmed fish have come to dominate many sectors
of the market. You've no doubt noticed a wide variation in
the price of salmon, from very inexpensive farmed salmon
to very expensive, fresh Alaskan salmon. Many environmental groups are opposed to farm-raised salmon and there
is some controversy about their omega-3 content, as they're
not always fed the marine diet that produces high amounts
of omega-3 fatty acids. In my opinion, the best salmon is
U.S. Pacific wild Alaska salmon, whether it's fresh, frozen,
or canned. The Marine Stewardship Council certifies
Alaska salmon as a "Best Environmental Choice."

Other heart-healthy, environmentally safe seafood
choices include the following: Arctic char, catfish (U.S.
farmed), clams (farmed), crab (Dungeness), crayfish, halibut (Alaskan), herring, mahi mahi, mussels (farmed),
sablefish, sardines, scallops (farmed), striped bass, and
tilapia (farmed).

Due to the unresolved issues relating to potential environmental toxins in some farmed salmon and the seemingly real threat some salmon farms are to the environment, we cannot at this time recommend farm-raised salmon (whether canned, frozen, or fresh).

To be environmentally responsible in your fish choices, consult the following websites for the latest information. Environmental issues surrounding shrimp are particularly complicated and a look at these websites will help you be an informed consumer:

www.audubon.org (888-397-6649)

www.environmentaldefense.org (202-387-3525)

www.mbayag.org (831-648-4800)

Most Americans don't eat enough seafood—or enough of the right kind of seafood (fried shrimp does not count!). The reason is obvious: it's not always easy to find good, fresh fish locally. Some of us have great local seafood markets; others are miles from any fresh fish outside a pet store. Here are two solutions: canned wild Alaskan salmon or canned Albacore tuna and/or frozen fish.

Canned wild Alaskan salmon can sit in your pantry for months. Canned sockeye salmon has 203 milligrams of calcium—17 percent of your daily requirement—as an added bonus if it's canned with the bones; don't worry, the fish has been cooked and the bones are so soft as to be unnoticeable. You can add the salmon to a green salad for a delicious light meal. You can make salmon-burgers out of it that are irresistible. Canned tuna is another good choice (although without the calcium boost). Just be sure to buy albacore tuna packed in spring water. Canned sardines are another excellent source of beneficial marine-derived fatty acids, vitamin D, plus the beneficial calcium from the hidden soft bones. Select sardines packed in tomato sauce for the added benefit of lycopene or soybean oil or olive oil. If you're a beginner at sardines, try the ones packed in olive oil; they have the best taste in my opinion.

∙∙

Consumer Action Alert

Ask manufacturers to produce canned fish with less sodium! You can always add a little salt if your taste buds demand it.

∙∙

Frozen fish can be an excellent alternative to fresh. Many stores—Trader Joe's and Whole Foods—make a point of offering environmentally safe, high-EFA frozen fish. Just be sure to defrost it slowly in the refrigerator, to preserve texture and flavor.

Of course, fresh wild salmon, trout, or sea bass is also terrific. Get to know your fishmonger and don't be shy about asking which is the freshest fish he has available.

∙∙

Don't think you can get your family to eat fish two or three times a week? Don't give up. As little as one serving of fish a week resulted in a significantly reduced risk of total cardiovascular deaths after eleven years in a study involving more than twenty thousand male physicians in the United States. Some is better than none!

∙∙

GRILLED WILD ALASKAN SALMON BURGERS

MAKES 4 SERVINGS

From the Alaska Seafood Marketing website
    One 14¾-ounce can wild Alaska salmon
    2 tablespoons lemon juice
    1½ tablespoons Dijon mustard
    ¾ cup dried bread crumbs

½ cup sliced green onions
Two omega-3-enriched eggs

Drain and flake the salmon. Combine the lemon juice and mustard. Blend the flaked salmon with the bread crumbs, green onions, and lemon juice–mustard mixture. Mix in the eggs until well blended. Form the mixture into 4 patties (chilling in the fridge for an hour will help them hold their shape) and cook on a lightly oiled grill or sauté in a skillet until golden brown on both sides. Serve each burger on a whole wheat bun with lettuce, tomato slices, and condiments as desired.

• •

An excellent source for seafood by mail is www.vitalchoice seafood.com.

• •

• •

For Those Who Just Won't Eat Fish . . .

Every now and again I run into a patient who tells me that he simply can't/won't eat fish. Ever. If that describes you, then take in supplement form at least a gram of EPA/DHA per day with food. If I'm not having any omega-3s any day, I take a supplement containing 500 milligrams of EPA/DHA with two of my meals. If you do take fish oil capsules, also take on a daily basis 200 to 400 IU of vitamin E. Look for brands of fish oil that list a small amount of d-alpha tocopherol (vitamin E) on the label. It keeps the fish oil fresh. Store the capsules in the refrigerator after opening. I use Trader Joe's brand, Trader Darwin's Omega-3 Fatty Acid Dietary Supplement. Any "fishy burps" you experience will diminish after three or four days of supplementation.

• •

See the SuperFoods Rx Shopping Lists (pages 329–30) for some recommended wild salmon foods.

Soy

Soy contains:

- Phytoestrogens
- Plant-derived
  omega-3 fatty acids
- Vitamin E
- Potassium
- Folate
- Magnesium
- Selenium
- Excellent nonmeat
  protein alternative

One of the national morning shows recently featured a cooking segment about the nutritional benefits of tofu.

"This will be great," said the host of the show. "I'm trying to get more soy into my diet."

"Well, then, this should work out really well," replied the co-host. "You can have mine!"

This brief humorous interchange typifies the way many of us think about soy and tofu, in particular. We think we should eat more of it, though we may not be sure why, and some of us are convinced we want no part of it whatsoever.

My goal in the next few pages is to convince you that soy is a valuable addition to your diet and, even if you never dreamed you'd eat it and even if you never, ever want to cook tofu, there are other ways to incorporate soy foods into your daily diet.

Here is the good news in a nutshell: soy truly is a Super-Food. It offers tremendous health benefits when incorporated into your diet. It's an inexpensive, high-quality, vitamin- and mineral-rich plant protein with lots of soluble fiber, plant-based omega-3 fatty acids, and, most important, it offers a wealth of disease-fighting phytonutrients. Indeed, soy is the richest known dietary source of powerful health-promoting phytoestrogens. Soy has been recognized by many researchers as playing a positive role in preventing cardiovascular disease, cancer, and osteoporosis as well as helping to relieve menopausal and menstrual symptoms. Moreover, you don't have to eat tons of it to enjoy its considerable advantages. Once you learn about the proven benefits of soy and the simple ways you can incorporate this unique food into your diet, I think you'll become a convert.

Soybeans have been cultivated in China since the eleventh century B.C. Indeed, the soybean is the most widely grown and utilized legume in the world. The Chinese name for the soybean is "greater bean," and soy is also referred to as "meat without bones." Like other beans, soybeans grow in pods, and while we most commonly think of them as green, they can also be yellow, black, or brown. The soybean was introduced to America in the eighteenth century by that innovative, forward-looking American Ben

Franklin, who, impressed with tofu—the Chinese "cheese made from soybeans"—had some beans shipped from Paris to a group of farmers in Pennsylvania. It wasn't until the next century that soybeans were extensively planted by American farmers. In the twentieth century, people began to recognize the health-promoting qualities of the soybean, and today, to many people's surprise, the United States is the world's largest commercial producer of soybeans.

SOY, THE BLANK CANVAS

One of the most unusual aspects of soy, at least in comparison with other SuperFoods, is also perhaps its great advantage: *you can make it taste however you like.* You can use it in a wide variety of preparations, and if you're not wild about tofu, you may well enjoy roasted soy nuts. You can add a scoop of soy protein to shakes or pancakes or bake with soymilk and it's undetectable. My own children had been eating soy for years before they even heard the word. The message is that you're sure to find some version of soy that you can happily live with. Soy foods all come from soybeans, but there are many permutations of the basic bean as I'll detail later.

••

Soy and Lactose Intolerance

Many people have difficulty digesting lactose—the main type of sugar found in dairy products. People with lactose intolerance experience upset stomachs and diarrhea when they consume dairy foods, particularly milk. Fortunately, soy foods allow people with lactose intolerance to consume the required protein and calcium they need without difficulty.

••

SOY: THE ALTERNATE PROTEIN

Before we even consider soy's health-promoting abilities, let's take a brief look at its other, often overlooked plus: it is an excellent protein alternative. For example, a half-cup of tofu provides 18 to 20 grams of protein, which is 39 to 43 percent of a daily requirement for adult women. That same amount of tofu also provides 258 milligrams of calcium (more than a quarter of our daily needs) and 13 milligrams of iron (87 percent of a woman's daily need and 130 percent of a man's). Here's a comparison of the percentage of protein by weight of a few foods: soy flour is 51 percent protein; whole, dry soybeans are 35 percent protein; fish is only 22 percent protein; hamburger is only 13 percent protein; and whole milk is just 3 percent protein.

Substituting 15 grams of soy protein for 15 grams of animal protein would cause the current U.S. dietary ratio of animal-to-plant protein to fall from two to one to a more desirable one to one, the ratio it was in the early 1900s. At this level of intake, soy protein would still represent less than 20 percent of the average protein intake of U.S. adults.

In addition to the high-quality protein you get when you substitute soy for animal protein, you get a bonus of vitamins, minerals, and a good dose of phytonutrients. Soy has a healthy mix of fats and no cholesterol. In one study, the substitution of soy for animal products reduced coronary artery disease risk in the study subjects because of their subsequent reductions in blood lipids (such as LDL), homocysteine, and blood pressure. For those of us eating a typical American diet, what this means is that soy is so good because much of what we eat is so bad for us! Many of our protein sources come with additional less-than-desirable components, particularly saturated fat, as well as hormones, pesticides, and other negatives. Tofu is even low in calories compared with other plant-based protein sources. In fact, tofu has the lowest known ratio of calories to protein in any plant food except mung bean and soybean sprouts.

Soy offers the highest-quality protein of any plant food. Available in organic forms (and therefore free of any pesticides or other additives), it offers all nine essential amino acids and is a good source of plant-derived omega-3 fatty acids. So even if you only relied on soy as a meat substitute a couple of times a week, you'd be ahead of the game.

SOY AND YOUR HEALTH

Soy has long been recognized as a highly nutritious food. Western scientists became particularly interested in soy when they noticed that people eating Asian diets enjoyed lower rates of heart disease as well as less cancer and osteoporosis, and had fewer hormonal problems than those eating a typical Western diet. While much research still has to be done, there is now broad agreement on various connections between soy and health promotion.

Soy's most conclusively demonstrated benefit concerns cardiovascular health. There have been extensive studies on the cholesterol-lowering effect of soy. One frequently cited study, published in the *New England Journal of Medicine* in 1995, describes an analysis of thirty-eight different studies. The authors found that consumption of soy protein resulted in significant reductions in total cholesterol (9.3 percent), LDL cholesterol (12.9 percent), and triglycerides (10.5 percent) with a small though not significant increase in HDL cholesterol. A recent study (March 2003) in the *Journal of Nutrition* demonstrated that the intake of soy foods among the premenopausal women subjects was inversely related to their risk for coronary artery disease and stroke as well as other disorders. Similar studies have demonstrated the same effect with people with diabetes and people with high cholesterol.

•••

Isoflavones in Soy Foods

The USDA, in collaboration with Iowa State University, has compiled a listing of the isoflavone content of soy foods. The values are expressed in milligrams per single serving of the food. The foods are organized from the most isoflavones to the least. (From *Wellness Foods A to Z*)

	Calories	Fat (g)	Isoflavones
Soybeans, dried, cooked (1 cup)	298	15	95
Soybean sprouts (¼ cup)	171	9.4	57
Soy nuts (¼ cup)	194	9.3	55
Tempeh (4 ounces)	226	8.7	50
Soy flour, full fat (⅓ cup)	121	5.7	49
Tofu, firm (4 ounces)	164	9.9	28
Soymilk (1 cup)	81	4.7	24
Edamame, cooked (4 ounces)	160	7.3	16

•••

No one is precisely certain how soy lowers cholesterol, but the evidence of its doing so is so incontrovertible that in October 1999 the FDA gave soy its official backing by allowing soy-food manufacturers to make health claims on their packages. They are able to claim that soy protein, when included in a diet low in saturated fat and cholesterol, may reduce the risk of coronary heart disease by lowering blood cholesterol levels.

Soy has also been shown to be a potent cancer-fighting food. Various components of soy have demonstrated anti-carcinogenic effects. They include protease inhibitors, phytosterols, saponins, phenolic acids, phytic acid, and isoflavones. Two of the isoflavones in soy—genistein and daidzein—are worthy of particular attention because soy foods are their primary dietary source. These two isoflavones act like weak estrogens in the body. While their effects aren't completely understood, we do know that they can compete with stronger naturally occurring estrogens and in this way help prevent hormone-dependent cancers like those of the breast and prostate. The isoflavones bind to sites on cell membranes that would normally be inhabited by hormones that can stimulate the growth of tumors. In addition to blocking the action of potent, naturally produced hormones, genistein can also inhibit the activity of enzymes that encourage the growth of blood clots and tumors. While there have been some variations in study results linking soy intake to breast cancer reduction, epidemiological studies show that women in Southeast Asian populations who consume diets high in soy protein (10 to 50 grams a day) have a four to six times reduced risk of breast cancer compared with American women who normally consume minimal amounts of soy.

There is some controversy about the effects of soy in the diet and most of the controversy surrounds soy and breast cancer. There have been continuing studies on the role of soy in the diet of women who have been diagnosed with breast cancer and whether soy will stimulate or reduce tumor growth in such women. Because of this controversy, I take the most conservative approach in recommending soy. (If you have a history of breast cancer, please consult your health care professional about the pros and cons of soy intake.) It's one of the reasons I never recommend taking soy supplements—just whole soy foods. I can also report that a very recent, very extensive (January 2003) study in *Nutrition Review* has confirmed the safety of soy dietary isoflavones in the diet.

There were reports, eventually disproved, that soy led to an increase in the development of senility. Indeed, most populations at the higher end of soy consumption have lower rates of dementia than populations who do not consume soy.

• •

A study in *Nutrition and Cancer* reports that people who regularly consume as little as one and a half servings of soymilk daily enjoy better cancer protection than those who occasionally consume soy. Try to use soy daily, either as soymilk on your cereal or oatmeal or as soy protein powder in a fruit smoothie or as a snack of soy nuts. Studies suggest that two separate soy foods a day, in separate meals, work best.

• •

Does soy help with menopausal symptoms? While it's controversial, some evidence says it can. For example, researchers from the University of Bologna, in Italy, gave two groups of menopausal women 60 grams of either soy protein or a look-alike placebo of dried milk protein daily for twelve weeks. The women eating soy protein experienced significantly fewer hot flashes and night sweats than the placebo group. Again, soy's estrogen-mimicking isoflavones are responsible. As a woman's natural levels of estrogen fall during menopause, the isoflavones seem to help take up the slack.

There is evidence that, because of its estrogenlike behavior, soy contributes to bone health and thus helps stave off osteoporosis. One study of sixty-six postmenopausal women at the University of Illinois found that soy protein added to the diet significantly increased both bone mineral content and density in the spine after six months.

Many people are confused about how much isoflavones to consume per day. Estimates of dietary intake in people in Asian countries indicate that intakes of isoflavones from soy foods range from 15 to 50 milligrams a day. The average intake is 30 to 32 milligrams. From whole foods not fortified products.

Here's a quick breakdown of the major components of soy and how they promote health:

Isoflavones: Soybeans are the best-known source of these compounds, which act like antioxidants as well as estrogens. Two of the isoflavones in soy—genistein and daidzein—reduce the risk of coronary heart disease, mitigate hormone-related cancers, and decrease the ability of tumors to grow new blood vessels. Preliminary evidence suggests that genistein decreases the growth of new blood vessels in the retina, which can lead to vision loss from age-related macular degeneration.

Lignins: Bind with carcinogens in the colon, speed up the transit time of these carcinogens, thus reducing their potential negative effects, and scavenge free radicals.

Saponins: Phytonutrients that boost the immune system and fight cancer.

Protease Inhibitors: Block the activity of cancer-causing enzymes called "proteases" and thus reduce the risk of cancer. Protease inhibitors have been reported to suppress carcinogens.

Phytic Acid: Antioxidant that binds with and eliminates metals that can promote tumors.

Phytosterols: Nondigestible compounds that reduce cholesterol absorption in the bowel and may help prevent colon cancer.

Protein: Only plant-based complete high-quality protein, totally cholesterol free, and low in fat.

Oil: Healthy oil that is free of cholesterol and offers a beneficial ratio of fatty acids (low in bad fat; high in good fat). It's a source of plant-derived omega-3 fatty acids as well.

••

Soy sauce is not a good source of soy! It has little to no nutritional benefit and is very high in sodium.

••

SOY IN THE KITCHEN

As mentioned, soy is available in a wide variety of foods. Soybeans can be eaten whole—fresh or frozen as in edamame or dried as in soy nuts. They can also be fermented to make tempeh, miso, or soy sauce, of which the latter two are used primarily to flavor various sweet and savory foods. They can be soaked, mashed, and heated to create soymilk or curdled to make tofu or bean curd. They're processed to make oil, flour, and soy noodles. The key thing to remember about soy is that, while all soy foods are derived from the soybean, they are, except for soybeans themselves, essentially processed foods. This isn't bad; but it does mean that you should read labels when you choose soy foods, and you will have to do some experimenting to find the soy foods you like best.

A daily intake of 25 grams of soy protein is ideal. Here are some rich sources:

Four ounces of firm tofu contains 18 to 20 grams of protein.
One soy "burger" includes 10 to 12 grams of protein.
An 8-ounce glass (1 cup) of Edensoy original formula soymilk contains approximately 11 grams of protein.
One soy protein bar delivers 14 grams of protein.
One-half cup of tempeh provides 16 to 19 grams of protein.
One-quarter cup of roasted soy nuts contains approximately 15 grams of protein.

Here's the key to shopping for soy foods: **check the protein content on the label**. Many people get very confused about buying soy because they try to rely on the "isoflavones" content as listed on the label. Some foods don't list isoflavones. Some foods list isoflavone amounts that are not accurate. Some foods list isoflavone fortification, but I don't recommend relying on added isoflavone in food. There simply isn't evidence to confirm the long-term safety of isoflavone-fortified products.

In general, the best way to learn the isoflavone content of a food is to rely on the listed protein content. The protein content of the food is closely linked to the isoflavone content. You can get the benefits of soy with as little as 10 grams of soy protein a day. For example, ¼ cup of soy nuts has 15 grams of soy protein. While soy nuts are high in calories, most people love them and are delighted to eat a scant ¼ cup while relaxing at the end of the day. That's all it takes to get the benefit of soy!

Of course, there are many other soy sources. Here are a few:

Tofu: Tofu is perhaps the best-known version of soy. It's a white, cheeselike food made from curdled soymilk, which has been shaped into blocks. Tofu is available in a few varieties: firm, extra firm, soft, and silken. Firm and extra-firm tofu are good for slicing into stir-fries and soups. They can also be grilled or baked. Silken tofu is wonderful for using in smoothies or in dips, dressings, and toppings.

• •

Four Easy Ways to Get Soy into Your Day

1 cup soymilk on cereal
1 ounce soy protein powder in a fruit shake
¼ cup soy nuts as a snack
Dried cereals and breads containing soy

• •

Soymilk: Soymilk is a major source of soy protein. Soymilk is made from soybeans that have been finely ground, cooked, and strained. It comes with various additives and in a variety of flavors. It's widely available in aseptic packages, which keep for a long time and don't need to be refrigerated until opened. As Lorna Sass says in her excellent *The New Soy Cookbook*, ". . . not all soymilks are created equal. Tastes ranged from light, fresh and pleasantly sweet to musky, chalky, oily and intensely 'beany.' Color ranged from creamy white to dark caramel, with lots of shades in between." You really have to experiment with locally available brands to find a soymilk that pleases you.

It's critical that you read the labels on soymilk. The amounts of protein and calcium, as well as other vitamins and fat and sugar, vary considerably from one brand to another. Most brands have 6 to 11 grams of protein per 8 ounces (1 cup). My favorite soymilks are Westsoy Unsweetened Vanilla Organic Soymilk and Original Edensoy Extra Organic Soymilk. Keep in mind that soy fat is good fat. I personally like full-fat soymilk. Beware of added calories in the form of sugar.

Soymilk can be substituted for cow's milk in baking. Some people—particularly children who are used to cow's milk—may balk at the color and flavors of some brands of soymilk and therefore won't enjoy it on cereal, but if your family resists drinking it straight, try it in pancakes, cakes, muffins, etc.

Soynuts: Soy nuts are soybeans that have been soaked in water and baked or roasted until they're lightly browned. High in protein, isoflavones, and soluble fiber, they are also high in calories, so you have to limit your consumption to a reasonable amount, say, ¼ cup a day. (Remember that ¼ cup soy nuts has 136 calories.) While soy nuts are available in a honey-roasted form as well as various flavored and spiced versions, I recommend that you eat soy nuts with no added ingredients. Be sure to read the label and avoid soy nuts with any added oil and avoid added salt. Soy nuts make a great portable snack. I toss a few on my granola. One-quarter of a cup has about 15 grams of protein.

Edamame: Edamame are green soybeans still in their pods. Ideal because they're a whole food, they are available in the frozen food section of natural food markets and many supermarkets. Boil the pods in lightly salted water for a few minutes, then pop them right from the pods into your mouth. Edamame taste like slightly sweet lima beans. You can also find shelled soybeans frozen in bags, and these are great to add to soups, pasta sauces, salads, and stews. One cup of shelled edamame has about 23 grams of protein.

Soy Protein Powder: There are two kinds of soy protein powder and, I'll admit, it can be quite confusing when shopping for this popular additive to shakes and baked goods.

Soy protein concentrate comes from defatted soy flakes. It contains about 70 percent protein, while retaining most of the bean's dietary fiber. Depending on how the concentrate is prepared, it may or may not contain a significant amount of isoflavones.

When protein is removed from defatted flakes, the result is soy protein isolates, the most highly refined soy protein. Containing 92 percent protein, soy protein isolates possess the greatest amount of protein of all soy products. They are a highly digestible source of amino acids (building blocks of protein necessary for human growth and maintenance). Soy protein isolates are the substance commonly used in many soy–heart disease research studies.

Whichever soy protein powder you choose, be sure that it's not fortified with extra soy isoflavones.

Soy Flour: Soy flour has been processed from whole ground soybeans. Use it to increase the protein content of breads, cakes, and cookies. Soy flour contains no gluten, so it cannot be used to replace the wheat flour in baking. But you can use it to supplement your other flour: in yeast-raised breads: use 2 tablespoons of soy flour per cup of wheat flour; with quick breads, you can replace up to one-quarter of the wheat flour with soy flour. You may notice that breads made with soy flour brown more quickly than those made with just wheat. One-quarter cup of soy flour has 8 to 12 grams of protein.

If you're new to soy, it's well worth checking out some sources that I'm sure you'll find informative and inspiring. Look for *Amazing Soy* or *The Joy of Soy* by Dana Jacobi. Both books are filled with inspiring recipes and excellent general information on incorporating soy into your diet. Two other books on cooking with soy include *This Can't Be Tofu* by Deborah Madison and *The New Soy Cookbook* by Lorna Sass. Both will change the way you think of cooking with soy.

Also, take a look at some websites for more helpful information on soy in your diet:

soyfoods.com
soybean.org
soyproducts.com

Tempeh: Tempeh is a soy food made from soybeans that have been cracked and inoculated with a beneficial bacterium. It is fermented and then formed into flat blocks. Sometimes grains like brown rice, barley, or millet are added. Tempeh has a meaty taste and is often used as a meat substitute in cooking. It can be marinated and grilled as well as added to stews and pasta sauces. High in protein, fiber, and isoflavones, it is usually found in the refrigerated dairy section of your natural food store or supermarket. Tempeh can be frozen and, once defrosted, must be refrigerated. It will keep for about ten days. Three ounces of tempeh, or about ½ cup, has approximately 16 ounces of protein.

Miso: Miso, like tempeh, is a fermented soy food. There is a wide range of misos available, particularly if you search in Asian markets. Generally a strong-tasting, salty condiment, miso is perhaps most familiar as miso soup. It does provide soy isoflavones but, like soy sauce, its sodium content is high and thus doesn't make a good general source of soy protein.

SOY SUPPLEMENTS

People often ask me if they can just take a soy supplement. Once the health benefits of soy became well publicized, it wasn't long before soy isoflavones were available in health food stores. These supplements, containing concentrated soy isoflavones, are promoted as being beneficial particularly to women as relief from menopausal symptoms. You know my answer: if you want to get the health benefits of soy, you must rely on the whole food. For one thing, no one is certain what's in various supplements: they may contain more or less isoflavones than they claim. Moreover, it's not clear if soy isoflavones behave exactly the way they do when they arrive in our bodies via the whole food. My recommendation is to avoid them.

• •

Kitchen Tofu Tip

Many people are unaware that tofu, like milk or meat, is perishable. It won't last forever in the fridge. Pay attention to the expiration date on the package when you buy it (look for the latest date as you would on milk), keep it refrigerated at all times, and change the water in the package daily. Moreover, remember that tofu, like meat, can host unfriendly bacteria like salmonella. Make sure you work with it on a clean surface and wash prep surfaces (and your hands) with soap and water before and after handling.

• •

Check the SuperFoods Rx Shopping Lists (pages 350–51) for some recommended soy foods.

Spinach

SIDEKICKS: kale, collards, Swiss chard, mustard greens, turnip greens, bok choy, romaine lettuce, orange bell peppers
TRY TO EAT: 1 cup steamed or 2 cups raw most days

••

Spinach contains:

- A synergy of multiple nutrients/phytonutrients
- Low in calories
- Lutein/zeaxanthin
- Lutein/zeaxanthin
- Beta-carotene
- Plant-derived omega-3 fatty acids
- Glutathione
- Alpha lipoic acid
- Vitamins C and E
- B vitamins (thiamine, riboflavin, B_6, folate
- Minerals (calcium, iron, magnesium, manganese, and zinc)
- Polyphenols
- Betaine

••

It's very simple: you must eat your spinach. Along with salmon and blueberries, spinach is right up at the pinnacle of the *SuperFoods Rx* powerhouse choices. Spinach has more demonstrated health benefits than almost any other food. Is this because spinach is really one of the best foods in the world? Yes and no. Yes, because it is an incredibly nutritious food with a stunning roster of benefits. No, only because there may be other foods—particularly other dark green leafy vegetables like kale and collards—that are comparably nutritious. We have much more information on the benefits of spinach than any other potential candidate. Long recognized as a nutritional standout, spinach has been the subject of countless impressive studies.

There are a number of studies demonstrating an inverse relationship between spinach consumption and:

- Cardiovascular disease, including stroke and coronary artery disease
- A host of cancers including colon, lung, skin, oral, stomach, ovarian, prostate, and breast cancers
- Age-related macular degeneration (AMD)
- Cataracts

....

Superstar Sidekicks

Most of the SuperFoods have sidekicks, but in the case of spinach, the sidekicks listed on page 164 are powerful foods. Each of the green leafies offers a tremendous nutrient boost. Vary your green leafie intake among all the sidekicks and have at least two servings most days. Remember, a serving is 1 cup raw or a ½ cup cooked.

....

What is it about spinach and the other green leafies that makes them such powerful health promoters? In the old days, nutritionists would have pointed to one or two of the nutrients in spinach that elevated them to the top ranks.

With spinach it was iron. Remember Popeye? It was supposed to be the iron that made him such a powerhouse. He wouldn't dare take on Bluto without popping a can. But it's almost as if nutritionists were working with an 8-pack of crayons; today we're looking at the 250-pack. Based on what we know and are learning daily about micronutrients, we understand that it's the *synergy* of the wide range of all the nutrients and phytonutrients in green leafies that make them superstars.

••

CoQ_{10}

A member of Dr. Lester Packer's antioxidant network, one of the world's foremost antioxidant research scientists, coenzyme Q_{10} works in synergy with vitamins C, E, and glutathione. It's a key player in our skin's antioxidant defense mechanism against sunlight damage and also a significant player in mitochondrial energy production. (The mitochondria are the cells' energy factories.) Spinach is an important source of this critical antioxidant.

••

Although I've listed the most significant nutrients in spinach in the beginning of this chapter, here is a more complete list of everything we know thus far that's found in this SuperFood:

- The carotenoids lutein, zeaxanthin, and beta-carotene
- The antioxidants glutathione, alpha lipoic acid, and vitamins C and E
- Vitamin K (spinach is a major source of vitamin K)
- Coenzyme Q_{10} (spinach is one of the only two vegetables with significant amounts of it; the other is broccoli)
- The B vitamins (thiamine, riboflavin, B_6, and folate)
- Minerals (calcium, iron, magnesium, manganese, and zinc)

- Chlorophyll
- Polyphenols
- Betaine
- Plant-derived omega-3 fatty acids

This list, which, as I said, seems to be growing as we learn more about spinach, is truly formidable. With most SuperFoods, there are one or two nutrients in particular that push an individual food to best in category; with spinach, the list is so long and impressive that the wide range of individual nutrients coupled with the unmatched synergy of those nutrients make it a top SuperFood.

••

Betaine

This is a nutrient you will be hearing more about. Betaine is a derivative of choline, an essential fat, and it plays a role in homocysteine metabolism. Betaine supplementation has been shown to lower homocysteine levels in humans—an important step in lessening the risk of cardiovascular disease. The combination of dietary folate and betaine may be the best way to lower homocysteine. Great sources of betaine are spinach, wheat germ, oat bran, wheat bran, and whole wheat bread.

••

SPINACH IN YOUR EYES

Of all the chronic diseases spinach combats, the ones affecting the eye are of particular interest to me as an ophthalmologist. My mother, who was remarkably healthy until she passed away at age 91, suffered from age-related macular degeneration (AMD). At age 75 she was declared legally blind. For the last sixteen years of her life, despite her robust health, she was unable to enjoy life fully. She couldn't read, drive, watch TV, sew, or see a movie. Watching my mother during the last years of her life had a major influence on the direction of my work. It inspired me to want to learn everything I could about nutrition and chronic disease.

The macula of the eye is responsible for central vision—the type we need for close work like writing and sewing as well as for distinguishing distant objects and color. Sadly, as many as 20 percent of all 65-year-olds show at least some early evidence of age-related macular changes. By age 90, about 60 percent of Caucasians will be affected by AMD. Worse yet, there is no effective treatment that restores 20/20 vision. That leaves us with prevention, and one of the best sources of prevention is certain foods, particularly spinach and its sidekick green leafies, which along with consumption of dietary marine-based omega-3 fatty acids can offer real hope.

....

The macular pigment of the eye protects against age-related macular degeneration (AMD). The lower the macular pigment level, the higher the risk of AMD. The best foods to elevate macular pigment are spinach, kale, collards, and turnip or mustard greens, as well as yellow foods, such as corn, egg yolks, and orange bell peppers.

....

While no one is precisely certain what causes macular degeneration, there's ample evidence that free-radical damage from long-term exposure to light and ultraviolet radiation may play a role. We know for sure that cigarette smoking is a proven risk factor for AMD and most likely cataracts as well. Indeed, smoking is the most preventable cause of AMD.

Enter the two powerful carotenoids in spinach: lutein and zeaxanthin. A number of studies have shown an inverse relationship between dietary intake of foods rich in lutein/zeaxanthin and the incidence of AMD. A similar relationship has been found between dietary lutein/zeaxanthin and the prevalence of cataracts. We know for sure that as the lutein and zeaxanthin levels increase in the macula of the eye, there is a significant decrease in the amount of harmful light rays that reach the retinal cells that produce vision. There seems to be little doubt that the lutein/zeaxanthin provide protection.

Most of us involved in AMD research feel that long before we can see clinical evidence of AMD, adverse events are occurring in the retina. Preliminary data from my studies of people at high risk for later development of AMD are highly supportive of this hypothesis. Prevention of this devastating visual disability is most likely a lifelong job. The earlier you start, the better off your retina will be. At the same time, it is never too late to take action.

• •

The following people are at higher risk for AMD and cataracts:

- Women
- People with blue eyes
- Smokers
- Those with a history of cardiovascular disease and hypertension
- Obese people
- Anyone who spends a great deal of time outside in the sun
- People with a low intake of fruits and vegetables
- People who are farsighted (AMD risk only)

• •

Lutein and zeaxanthin also help prevent other eye maladies. Cataracts are a common occurrence in older people. Eighteen percent of people age 65 to 74 have cataracts and 45 percent of people age 75 to 84 have developed them. A cataract forms over the lens of the eye when, over time, damaged cells accumulate and cloud the lens. In one twelve-year Harvard study, 77,466 nurses over age 45 were found to demonstrate a clear relationship between lutein and zeaxanthin levels and the rate of cataract development. Overall, the nurses who ate the most dietary lutein/zeaxanthin had 22 percent fewer cataract surgeries. Another study of 36,000 male physicians had comparable results. Virtually every study on eye health has come to the same conclusion: the more lutein and zeaxanthin-rich food consumed—particularly spinach, kale, collards, and broccoli—the healthier the eye. I think of these powerful carotenoids as natural sunglasses for the eye. We also know that the chlorophyll in spinach is a potential cancer fighter. Preliminary studies suggest it may be helpful in preventing tumor cell growth and exerting a significant antimutagenic effect against a wide range of potentially harmful carcinogens.

Lutein and zeaxanthin are always found together in varying proportions in foods. There is preliminary evidence that zeaxanthin may exert an independent beneficial role in preventing macular degeneration. My recommended goal is 12 milligrams of lutein daily, which provides a variable amount of zeaxanthin. Optimum amounts of zeaxanthin are not yet known.

••

Lutein All-Stars

1 cup cooked kale	23.7 mg.
1 cup cooked spinach	20.4 mg.
1 cup cooked collards	14.6 mg.
1 cup cooked turnip greens	12.1 mg.
1 cup raw spinach	3.7 mg.
1 cup cooked broccoli	2.4 mg.

Zeaxanthin All-Stars

1 large orange bell pepper	8 mg.
1 cup canned sweet yellow corn	0.9 mg.
1 raw Japanese persimmon	0.8 mg.
1 cup degermed cornmeal	0.7 mg.

People often ask me why orange bell peppers are included as a sidekick to spinach, since it's not a leafy green. Orange bell peppers are extremely high in lutein/zeaxanthin. I tell my patients who hate spinach that they should eat orange bell peppers by cutting one up into a salad or adding it to a stir-fry. Most markets carry orange bell peppers all year round. I often put out a plate of cut-up peppers and baby carrots for the staff at my medical office and they disappear in no time.

The data I've presented on these peppers (see below) are preliminary and based on the most reliable published information available as well as on personal conversations with the holder of the patent for orange bell pepper seeds. The USDA should analyze this food, as it has not as yet included this nutrition powerhouse in its database.

One medium orange bell pepper contains:

0.4 mg. beta cryptoxanthin
1 mg. lutein
6.4 mg. zeaxanthin
0.3 mg. alpha-carotene
0.4 mg. beta-carotene
223 mg. vitamin C
4.3 mg. vitamin E

Another source of lutein/zeaxanthin, and one that's the most bioavailable of all, is the humble egg. While there's not a tremendous amount of lutein/zeaxanthin in egg yolks, what's there is so bioavailable that it's taken up into the bloodstream with great efficiency, giving a significant boost to the serum levels of these protective carotenoids. Eggs are quite nutritious. They're a good source of vitamin B_{12}, riboflavin, selenium, vitamin A, and vitamin D as well as lutein/zeaxanthin. They have a high quality of protein, due to their good balance of amino acids. An egg a day for most people (at least those who don't suffer from very high cholesterol and/or diabetes) is a fine addition to a healthy diet.

It's important to buy high omega-3 eggs, as they make a considerable contribution to your healthful balance of fatty acids. Look for "high omega-3" or "vegetarian fed" or "high DHA omega-3" on the carton. Here's a comparison between a typical supermarket egg and an omega-3 enriched egg.

	1 large egg	Egglands Best egg
Calories	75	70
Protein	6.3	6
Total fat	5	4
Saturated fat	1.5	1
Cholesterol	213	180
Vitamin E	0.5	about 3.8

SPINACH AND VITAMIN K

Spinach is a rich dietary source of vitamin K—a vitamin that unlike other fat-soluble vitamins is not stored by the body in appreciable amounts and must be replaced on a regular basis. We're discovering more each day about the importance of this vitamin, and it seems the more we learn, the more diverse and critical these functions are. (Here's yet another argument for the *SuperFoods Rx* approach: a wide range of nutrient-dense foods.) What we know so far: vitamin K is essential for the production of six of the proteins necessary for proper blood coagulation. Blood simply won't clot properly without it. It's been hypothesized that vitamin K plays a role in vascular health. Early work in this area is promising, though more studies must be done. We do know that low levels of vitamin K have been linked with lower bone density and an increased risk of hip fracture in women and that a serving a day of spinach significantly reduces this risk. Just 1 cup of fresh spinach leaves a day gives you 190 percent of your daily requirements of vitamin K.

CARDIOVASCULAR DISEASE AND SPINACH

Spinach is a heart-healthy food. The rich supply of carotenoids and other nutrients helps protect artery walls from damage. The greens highest in carotenoids include spinach, beet and mustard greens, kale, collards, and turnip and dandelion greens. Just a half-cup of cooked spinach supplies 95 percent of my suggested daily intake of beta-carotene and 85 percent of my suggested daily intake of lutein/zeaxanthin. Usually, we think of beta-carotene as associated with the color orange, as in pumpkins or sweet potatoes, but in spinach the orange beta-carotene is masked by the dense green of the chlorophyll in the spinach leaves.

An excellent source of both vitamin C and beta-carotene, which your body may convert to vitamin A, these nutrients in spinach work together to prevent oxidized cholesterol from building up in blood vessel walls. A cup of fresh spinach leaves can provide you with a substantial amount of your daily requirement of vitamin A (via beta-carotene) and 11 percent of the adult female requirement for vitamin C, and 9 percent of the RDA for males.

••

Leafy Greens and Your Blood Pressure

An easy way to boost your intake of antihypertensive nutrients is to eat leafy greens. They're high in potassium and low in sodium. They provide calcium, magnesium, folate, polyphenols, fiber, and at least a trace to measurable amounts of plant-derived omega-3 fatty acids. Your blood vessels will thank you for this combination of nutrients.

••

Spinach is also an excellent source of folate. Folate plays a significant role in preventing cardiovascular disease because it works to escort a dangerous amino acid—homo-cysteine—from the body. We know that elevated levels of homocysteine are associated with increased risk of heart attack and stroke. Folate is also a key nutrient in DNA repair. This important B vitamin thus plays a major role in cancer prevention. The potassium and magnesium in spinach also contribute to cardiovascular health, as they both work to lower blood pressure and reduce the risk of stroke.

••

To enhance your absorption of carotenes, toss cooked greens with a teaspoon of healthy extra-virgin olive oil and/or some chopped nuts and avocado, or as a side dish with a piece of salmon.

••

SPINACH AND CANCER

In epidemiological studies, it's been found that the more spinach consumed, the lower the risk of almost every type of cancer. It's not surprising that spinach would be a powerful anticancer food given the high level of nutrients/phytonutrients it contains. There are a number of different flavonoid compounds in spinach working to prevent different stages of cancer development. Glutathione and alpha lipoic acid are two antioxidants that some researchers believe are the most important in the body. Normally these life-preserving nutrients are manufactured in the body itself, but our ability to produce them seems to diminish as we age. However, spinach contains a ready-made supply of both. Glutathione is the primary antioxidant in all cells where its critically important job is to protect our DNA. It repairs damaged DNA, promotes healthy cell replication, boosts the immune system, detoxifies pollutants, and reduces chronic inflammation. Alpha lipoic acid not only boosts glutathione levels, it helps stabilize blood sugar. Studies suggest it has an antiaging role (e.g., a favorable influence on age-related mental decline) and helps prevent cancer, heart attacks, and cataracts. Alpha lipoic acid is unusual in that it's both fat and water soluble. It can work in the fatty part of cell membranes and also in the water portions of our cells to reduce oxidative damage.

Lutein, another powerful antioxidant in spinach, works to enhance the body's immune system, thus warding off many types of cancers. Greens seem to be particularly effective in preventing stomach cancer. A Japanese study found that a higher intake of yellow-green vegetables could cut the risk of gastric cancer in half.

• •

As a general rule, the darker the greens, the more bioactive phytonutrients they contain and thus the more powerful they are against cancer and other diseases.

• •

Former smokers in particular can benefit from the power of spinach. Studies have found that people who eat a serving of spinach or one of its sidekicks daily, even if they're former smokers, have a significantly reduced risk of developing lung cancer. There's little question that people who consume the least carotenoid-rich food like spinach roughly double their risk of developing lung cancer.

Calcium and Spinach

While spinach is relatively rich in calcium, the calcium it contains is bound to oxalates and is not readily bioavailable. However, the oxalates in spinach have a minimal effect on calcium absorption from other foods eaten with spinach. In other words, if you consume yogurt or any other calcium-rich food along with spinach, you'll still benefit.

THE LEAFY DIFFERENCE

Not all green leafies are created equal. Many of us are used to just grabbing a head of iceberg every time we think salad. If you expand your horizons, you'll get far more nutrition from your salads, as well as sandwiches, tacos, and all other dishes that call for something green and leafy.

Green Leafy Comparisons

(based on 1 cup raw of each)	Spinach	Romaine	Iceberg
Calories	7	9	6
Fiber	<1 g.	<1 g.	>1 g.
Calcium	30 mg.	18 mg.	11 mg.
Iron	0.8 mg.	0.6 mg.	0.2 mg.
Magnesium	24 mg.	8 mg.	4 mg.
Potassium	167 mg.	140 mg.	84 mg.
Zinc	0.2 mg.	0.1 mg.	0.1 mg.
Vitamin C	8 mg.	13 mg.	2 mg.

Green Leafy Comparisons

(based on 1 cup raw of each)	Spinach	Romaine	Iceberg
Niacin	0.2 mg.	0.1 mg.	0.1 mg.
Folate	58 mcg.	76 mcg.	31 mcg.
Vitamin E	0.6 mg.	0.1 mg.	0.2 mg.
Lutein/Zeaxanthin	3.7 mg.	1.4 mg.	0.2 mg.
Beta-carotene	1.7 mg.	2 mg.	0.1 mg.

As you can see, spinach is king of this crowd. It has the most total carotenoids. Romaine has only 38 percent of the lutein found in spinach but does have slightly more beta-carotene. Iceberg has only 5 percent of the lutein and 6 percent of the beta-carotene that spinach has and 17 percent of the magnesium and 53 percent of the folate. If you're used to iceberg-only salads, try mixing romaine and spinach. I often make a salad of one-half spinach to one-half romaine. You'll get the crispy crunch that you do from iceberg lettuce and lots more nutrients.

SPINACH IN THE KITCHEN

Spinach and many greens are available in markets all year round. There are different varieties, ranging from the Savoy type with crinkly, curly leaves to the flat- or smooth-leaf type, which has unwrinkled leaves. Spinach is sold both loose and in bags. Except for the bagged baby spinach, I prefer to buy loose spinach because it's easier to examine for freshness. It should always smell sweet and the leaves should be crisp and intact. Bagged greens can deteriorate quickly and should be carefully examined for darkened leaves that might signal they're past their prime. Yellowed leaves are also a sign that the greens are a poor choice. Spinach and most greens will only keep for three to four days after purchase. Don't wash spinach before storing, as that hastens deterioration. Wrap loose spinach in paper towels and store in the crisper.

Before spinach is cooked or served in a salad, it must be washed and washed! The leaves tend to harbor sand. Tear the leaves from the tough center stem—if using baby spinach, there's no need to do this—and put the leaves in a large bowl or sink filled with cool water. Allow the dirt to sink to the bottom, lift out the spinach, drain the water and sand, and repeat until all the grit is out. Don't soak the spinach; any greens will lose valuable vitamins if they're left to soak in water. A dip, swish, and rinse is the way to go.

The worst chore connected with spinach is the washing, rinsing, washing, rinsing, and so on. If you find great spinach at a roadside stand, it's worth it to go through the process, but at other times you're just in too much of a rush. My ship really came in when I found Ready Pac brand prewashed baby spinach in the supermarket. You can microwave it right in the bag after slitting it to let the steam escape. We go through several bags a week.

••

Cooked or Raw?

Greens are best eaten both raw and cooked. Cooking liberates the carotenoids, especially beta-carotene, and makes it more bioavailable. It also boosts lutein. But heat degrades both vitamin C and folate. The best approach? Enjoy greens in both salads and cooked form.

••

QUICK WAYS TO GET SPINACH AND OTHER GREENS INTO YOUR DIET

- Layer cooked spinach or other greens in a lasagne.
- Steam spinach and serve sprinkled with fresh lemon juice and grated Parmesan cheese. This keeps for three days in the fridge, so I enjoy it as leftovers or take some to work.
- Add a handful of spinach leaves to soups.
- Dress leftover greens with balsamic vinegar dressing and sprinkle with some sesame seeds.
- Add chopped greens to an omelet along with chives, tomato, bell peppers, and onion.
- Shred various greens along with romaine lettuce in a salad.
- Shred greens onto tacos and burritos.

Purslane

Many people consider purslane, if they consider it at all, a common weed. It's an annual that thrives in dry, sandy soil and can often be found by roadsides and even, perhaps, at the edges of your own garden. Purslane is actually a SuperFood. Long regarded—indeed, as far back as ancient times—as a remedy for heart problems, sore throats, swollen joints, dry skin, and a variety of other ailments, purslane was and is commonly eaten in Greece, Europe, Mexico, and Asia. Purslane is a worthy addition to any diet: it's actually the very best source of plant-derived omega-3 fatty acids, and a good source of vitamin C, beta-carotene, and glutathione. Eaten in a salad with some olive oil and a bit of lemon juice, purslane is delicious. If you're interested in trying purslane, check websites or books for photos to identify it, and see if you can find some growing wild. Be sure not to pick any from land that might have been treated with chemicals. Look for purslane at farmers' markets, or try growing your own. It's easy to grow, resists drought, and self-seeds readily. A number of seed suppliers offer purslane seeds. Try Seeds of Change (www.seedsofchange.com), Eden Seeds (www.edenseeds.com), or Bountiful Gardens (www.bountifulgardens.org).

PATTY'S SPINACH PESTO

Puree raw spinach with almonds or walnuts, some garlic, olive oil, and Parmesan cheese. This is delicious on chickpeas or bow-tie pasta. It can be frozen.

Tea

SIDEKICKS: none
TRY TO DRINK: 1 or more cups daily

••

Tea contains:

- Flavonoids
- Fluoride
- No calories

••

How about a SuperFood that's cheap, has no calories, is associated with relaxation and pleasure, tastes good, and is available everywhere, from the finest restaurants to the local diner? And how about if that food lowered blood pressure, helped prevent cancer and osteoporosis, lowered your risk for stroke, promoted heart health, played a probable role in preventing sunlight damage to the skin (such as wrinkles and skin cancer), and contributed to your daily fluid needs? And what if, to boot, it were antiviral, anti-inflammatory, anticavity, antiallergy, and prevented cataracts? Tea is all that. If you're not sipping orange pekoe at the office, gulping refreshing brewed iced green tea on the tennis court, or enjoying some Earl Grey after dinner, you're missing an opportunity to improve your health and longevity with tea, the world's most popular SuperFood.

According to legend, the discovery of tea occurred quite by accident in 2700 B.C. in the reign of the Chinese emperor Shen Nung. As the emperor rested beneath a shade tree, a servant boiled some drinking water nearby. A breeze came up and blew some leaves from a nearby wild tea tree into the pot. The emperor, impatient to drink, sipped the water and was delighted with the taste. Thus was born a drink that is, after water, the most popular drink in the world. There are more than 3,000 varieties of tea available around the world, and it's a beverage that, because of its complexity and variety, attracts both connoisseurs and ceremony. From the British institution of tea time to formal Japanese tea ceremonies, no other beverage, save perhaps wine, inspires such ritual and debate.

While the savoring of tea's culinary attractions is an ancient pastime, the health-promoting properties of the beverage have recently drawn wide attention. Interest in the medicinal properties of tea has ebbed and flowed over the centuries, but it hasn't been until recently that research has confirmed ancient suspicions: tea—the simple, common beverage—is a healthy drink.

All true tea comes from a single plant: the evergreen *Camellia sinensis*. (Herbal teas are not considered true teas, but rather are beverages brewed from herbs, roots, and other sources. While some have medicinal properties, they are a separate category from authentic tea.) Three types of tea are produced from this single shrub: green, black, and oolong. The differences are in the way the leaves are processed after harvesting. Green tea is lightly processed. Favored in Japan, it comprises about 21 percent of the world's tea production. Black tea, favored in Europe and the West, makes up about 77 percent of tea production worldwide. Black tea is made of leaves that have been left to ferment following harvesting. This fermenting darkens the leaves and allows them to develop a stronger flavor. Oolong tea, popular in China and Taiwan, is partially fermented.

While green tea has received the lion's share of the attention regarding health benefits, in fact, all true teas are beneficial. Green tea has been studied more extensively.

Tea contains more than four thousand chemical compounds. The ones that have drawn the most attention, and which have proven benefits, include the phytonutrient polyphenols called "flavonoids"—the same type that is found in red wine and berries. There are about 268 milligrams of flavonoids in a cup of brewed black tea, and about 316 milligrams of flavonoids in a cup of green tea. One cup of brewed green tea provides more than five times the flavonoids than red onion, another popular flavonoid all-star. The most potent polyphenol in tea is a substance known as epigallocatechin gallate, or EGCG, which belongs to a group of flavonoid phytochemicals known as catechins. Research has shown that in a test tube the catechins are more effective antioxidants than even the powerful vitamins C and E. In one laboratory test, the EGCG in green tea was found to be twenty times more potent an antioxidant than vitamin C.

• •

One cup of black tea contains about 268 milligrams of flavonoids. One cup of green tea contains 316, but only half that amount if it's decaffeinated. You can significantly increase this amount by squeezing the bag after steeping for three to four minutes. Herbal-tea infusions are not a significant source of polyphenols.

• •

Which tea is best? While it used to be thought that green tea was the standout in promoting health, we now know that both green and black tea have similar, distinct and, in some cases, overlapping biochemical, physiological, and epidemiological effects. It's possible that there may be an occasional disease where one is more effective than the other, but overall the message is to choose the tea you like best and enjoy it. I drink both black and green tea at different times in the day.

There is some evidence that the health benefits of tea may be attributable to its caffeine content. Caffeine seems to have antimutagenic properties, which may be associated with an anticancer effect. Epidemiological studies suggest that caffeine may provide protection from the development of Parkinson's disease. So it would seem to be better to drink it in a caffeinated form. If caffeine is a problem for you, limit your tea drinking to early in the day and try green tea, which is generally lower in caffeine content.

••

There is less caffeine in tea than in an equal amount of coffee—roughly one-third less—and it seems to elicit fewer of caffeine's typical side effects.

••

Some of the research on the benefits of tea has been contradictory. For example, some of the evidence pointing to tea's effects has only been gathered in the laboratory. Will the positive results be duplicated in humans? We'll have to wait and see. In some instances, there have been negative associations with tea and health, but in many of these cases there have been other mitigating factors. Studies have shown tea to be both a positive and a negative influence on the development of esophageal cancer, but researchers speculate that the negative results could have been due to the way in which the tea was consumed. In some countries, tea is drunk boiling hot and/or is heavily salted, and either of these two preparations is known to encourage the development of cancers.

I believe the very positive news about tea is convincing. And, like the other SuperFoods, tea should be viewed as a part of a healthy lifestyle: you can't smoke, abuse alcohol, overeat, and never exercise, and expect tea to save you. And you can't think that you can rely on the polyphenols in tea and skip the fruits and vegetables. On the other hand, I believe that tea is a wise addition to a healthy lifestyle for its ability to promote health and prevent disease.

••

Have a cup of green or black tea before you exercise in the morning. The flavonoids begin to appear in your blood within about thirty minutes, giving you an antioxidant boost and thus preparing your body to handle the free radicals generated by exercise.

•••••••••••••••••••••••••••• ••••••••••••••••••••••••••••••

TEA AND CANCER

There is evidence that tea consumption decreases the risk of stomach, prostate, breast, pancreatic, colorectal, esophageal, bladder, and lung cancers. Laboratory studies have consistently shown that tea can inhibit the formation and growth of tumors. Researchers have demonstrated that the catechins in tea prevent cell mutation and deactivate various carcinogens. They also decrease the growth of cancer cells and inhibit the growth of the blood vessels that tumors need in order to grow. In one Japanese study, researchers showed that women who drank the most green tea—in some cases as much as ten cups daily—had a lower risk of developing cancer when compared to women who did not drink tea. There is also a belief among some researchers that the prostate-cancer incidence in U.S. males is fifteen times higher than the incidence in Asian males, in part because of the considerably greater amounts of tea drunk by Asians. It also seems that teas may possess a probiotic effect, which enhances gastrointestinal health.

While even one cup of tea seems to provide health benefits, it may take as many as four cups daily to really achieve a major decrease in cancer risk.

• •

While research on dementia and the consumption of tea has not been completed, we do know that people with the highest intake of flavonoids seem to have the lowest risk for developing dementia.

• •

TEA AND HEART HEALTH

There is solid evidence that tea consumption is associated with a lowered risk of heart disease and stroke. The connection was noticed when the arteries of Chinese-American tea drinkers were compared with the arteries of Caucasian coffee drinkers. The tea drinkers had only two-thirds as much coronary artery damage and only one-third as much cerebral artery damage upon autopsy compared with the coffee drinkers. Another study found that in males, deaths from coronary artery disease were reduced by 40 percent among those who drank one or more cups of tea daily, and another study from Harvard showed that there was a 44 percent lower risk of heart attack in people who drank at least one cup of tea daily.

While some studies on tea and coronary artery disease have been inconclusive, in animal studies we know for sure that the catechins lower cholesterol levels, especially the damaging LDL cholesterol. There's also a definite inverse relationship between tea consumption and homocysteine levels, which are of course associated with an elevated risk for heart disease. Tea also seems to play a role in keeping the lining of the blood vessels plaque free, which in turn lessens the risk of coronary artery disease. It seems that these positive benefits can be enjoyed if you drink between one and three cups daily, with greater protection conferred as the total consumption increases.

Interestingly, one study showed that tea consumption in the year before a heart attack is associated with a lower mortality following the heart attack. In this study, moderate tea drinkers drank less than fourteen cups weekly, compared to those who drank none and those heavy tea drinkers who drank fourteen or more cups weekly. Both the moderate and the heavy tea drinkers had a lower death rate than those who abstained entirely. The heartening implication of multiple studies is that one does not need to consume tremendous amounts of tea to enjoy health benefits. As little as a cup a day can play a positive role in your health.

• •

Preliminary data suggest that tea may actually help you lose weight by increasing energy expenditure.

• •

MORE GOOD NEWS

Tea seems to have a positive effect on your dental health. Drinking tea lowers your risk of developing cavities as well as gum disease. One study found that tea may reduce cavity formation by up to 75 percent. This happens for a number of reasons. The fluoride content of the tea inhibits cavities from developing. Tea also seems to inhibit bacteria from adhering to tooth surfaces, while it also inhibits the rate of acid production of oral bacteria.

There may be a positive association between tea consumption and the prevention of kidney stones. Although some publications suggest the reverse, in the Nurses' Health Study it was found that for every cup of tea consumed daily, the risk of developing kidney stones decreased by 8 percent.

Both men and women can improve bone health by drinking tea. Studies that focused on the risk of hip fracture found that habitual tea consumption, especially when maintained for more than ten years, has been shown to have a significant benefit to bone-mineral density. This seems to be due to the fact that some of the flavonoids in tea have phytoestrogen activity, which benefits bone health. Moreover, some tea extracts seem to inhibit bone resorption.

One recent study found that oolong tea is successful in treating atopic dermatitis; this is no doubt due in part to the anti-allergic properties of tea. This benefit was noticed after one or two weeks of drinking tea. In this study, a ⅓-ounce tea bag that steeped for five minutes in just over four cups of boiling water was consumed in three parts, one with each meal.

SOME TEA TIPS

- Brewed tea confers more health benefits than instant tea.
- Tea bags are as potent as loose tea in their health benefits.
- Brew tea for at least three minutes.
- Squeeze the brewed tea bag to almost double the polyphenol content.
- Add a wedge of lemon or lime with the rind for a polyphenol boost.
- If you're caffeine sensitive, reduce brewing time to one minute or so.

- Avoid drinking extremely hot tea.
- The flavonoids degrade with time, so it's best to drink freshly brewed tea that's hot or quickly iced.

See the SuperFoods Rx Shopping Lists (pages 351–52) for some recommended tea choices.

Tomatoes

SIDEKICKS: red watermelon, pink grapefruit, Japanese persimmons, red-fleshed papaya, strawberry guava
TRY TO EAT: one serving of processed tomatoes or sidekicks per day and multiple servings per week of fresh tomatoes

••

Tomatoes contain:

- Lycopene
- Low in calories
- Vitamin C
- Alpha- and beta-carotene
- Lutein/zeaxanthin
- Phytuene and phytofluene
- Fiber

- Potassium
- B vitamins (B_6, niacin, folate, thiamine, and pantothenic acid)
- Chromium
- Biotin

••

Many people feel certain that there's going to be bad news along with the good news when they learn that the tomato is a SuperFood. Of course, they like tomatoes, they think, but in many locations tomatoes are only tasty for a couple of months each year. Well, there's good news and more good news about tomatoes. Not only do they pack a nutritional wallop, but you can enjoy their benefits all year long. That's because their power is available in processed tomatoes. The spaghetti sauce and taco sauce that you love, along with that slice of pizza and even, yes, ketchup and barbeque sauce, all have the power of tomatoes. So, no matter where you live, it's easy to get more tomatoes into your diet and begin to enjoy their considerable benefits.

The tomato—a critical ingredient in some of our favorite foods, including pizza and lasagne—has had a checkered past. Once scorned as a sinister and poisonous food (one Latin name, *lycopersicon* or "wolf peach," refers to the belief that tomatoes were like a wolf—dangerous). It wasn't until the late nineteenth century that tomatoes became popular. Originally grown and enjoyed by the Aztecs in Mexico, tomatoes were imported to Europe by Spanish missionaries. Viewed as a dangerous food by all but the Italians and Spanish, it took years for tomatoes to lose their unsavory reputation.

There was some basis for the original skepticism that clung to tomatoes: their leaves do contain toxic alkaloids. Embraced by Americans by the end of the nineteenth century, tomatoes have gone on to become one of our most popular vegetables and now are recognized as one of our favorite SuperFoods.

It should be noted that tomatoes are not really vegetables. Botanically classified as a fruit, they are the seed-bearing portions of a flowering plant. However, in 1893, a case came before the Supreme Court of the United States relating to shipping tariffs on tomatoes. Should farmers pay fruit or vegetable rates on them? The Court came down on the side of vegetables, and so vegetables they became.

There are two new carotenoids on the scene with promising health benefits: phytoene and phytofluene are both found in tomatoes and tomato products. Phytoene has been shown to possess antioxidant capabilities and also anticarcinogenic action. More study needs to be done, but preliminary work indicates that they play a role in tomatoes' ability to combat cancer and other diseases.

THE POWER OF RED

Lycopene, a member of the carotenoid family and a pigment that contributes to the red color of tomatoes, is a major contributor to their health-promoting power. Lycopene has demonstrated a range of unique and distinct biological properties that have intrigued scientists. Some researchers have come to believe that lycopene could be as powerful an antioxidant as beta-carotene. We do know that lycopene is the most efficient quencher of the free-radical singlet oxygen, a particularly deleterious form of oxygen, and lycopene is also capable of scavenging a large number of free radicals.

Lycopene is a nutrient whose time in the spotlight has come. It's been the subject of great interest lately as more and more researchers have focused on the particular power of this nutrient. The attention began in the 1980s when studies started to reveal that people who ate large amounts of tomatoes were far less likely to die from all forms of cancer compared with those who ate little or no tomatoes. Many other studies echoed the positive findings about the effect of eating tomatoes.

It's not only cancer that the lycopene in tomatoes helps mitigate. Lycopene is an important part of the antioxidant defense network in the skin, and dietary lycopene by itself or in combination with other nutrients can raise the sun protection factor (SPF) of the skin. In other words, by eating tomatoes (in this case, cooked or processed tomatoes) you're enhancing your skin's ability to withstand the assault from the damaging rays of the sun. It acts like an internal sunblock!

Lycopene may also indirectly lower the risk for age-related macular degeneration by "sparing" lutein oxidation so that lutein can be transported to the macula in its unoxidized, protective form.

● ●

Lycopene in Food

(22 milligrams of lycopene is the ideal daily amount)

	Milligrams
Tomato puree (½ cup)	27.2
Tomato juice (1 cup)	22
R. W. Knudsen Very Veggie vegetable juice cocktail (1 cup)	22
Tomato sauce (½ cup)	18.5
Watermelon wedge	13
Tomato paste (2 tablespoons)	9.2
Watermelon balls (1 cup)	7
Ketchup (2 tablespoons)	5.8
Stewed tomatoes, canned (½ cup)	5.1
Pizza (3-ounce slice)	4
Tomato (fresh, medium)	3.2
5 cherry tomatoes	2.2
½ pink grapefruit	1.8

● ●

Perhaps you've heard of the nun study in which Dr. David Snowdon of the Sanders-Brown Center on Aging at the University of Kentucky assessed eighty-eight Roman Catholic nuns ranging in age from 77 to 98. The nuns with the highest blood concentrations of lycopene were the most able to care for themselves and complete everyday tasks. Overall, those with the highest levels of lycopene were 3.6 times better able to function in their everyday lives than those with the lowest levels. Most interestingly, no similar relationship between vigor and the presence of other antioxidants (such as vitamin E and beta-carotene) was found.

Lycopene is rare in foods, and tomatoes are one of only a few that are rich in this powerful antioxidant. Indeed, ketchup, tomato juice, and pizza sauce account for more than 80 percent of the total lycopene intake of Americans.

Red watermelon is another excellent source of lycopene. A very concentrated, bioavailable source of the nutrient, some food sources say that, ounce for ounce, watermelon is even richer in lycopene than tomatoes. Watermelon definitely results in a blood level boost of lycopene comparable to that of tomatoes.

Pink grapefruit has not been studied as extensively as tomatoes and watermelon, but the lycopene in this food is nonetheless substantial. When watermelon and pink grapefruit are eaten, the efficient absorption of the lycopene depends on the presence of a bit of dietary fat. Tomatoes are often served with olive oil or cheese, while pink grapefruit and watermelon are often served on their own. Just be sure that you eat them in conjunction with some fat; even a couple of nuts will do, or with pink grapefruit, have a slice of toast with some cheese or avocado. A salad of cubed watermelon and some feta cheese is refreshing and does the trick.

• •

Watermelon, strawberry guava, pink grapefruit, red-fleshed papaya, and persimmons are other dietary sources of lycopene. All make excellent additions to the diet.

• •

While lycopene has received a lot of attention recently, tomatoes are rich in a wide variety of nutrients, which seem to work synergistically to promote health and vitality. Low in calories, high in fiber, and high in potassium, tomatoes are not only a rich source of lycopene, they are also a source of beta-carotene, alpha-carotene, lutein/zeaxanthin, phytuene/ phytofluene, and various polyphenols. They contain small amounts of B vitamins (thiamine, pantothenic acid, vitamin B_6, and niacin), as well as folate, vitamin E, magnesium, manganese, and zinc.

It's the synergy of this multitude of nutrients, as well as the special power of lycopene, that boosts tomatoes to a spot in the all-star SuperFood pantheon.

TOMATOES AND CANCER

Some of the most exciting studies on tomatoes have focused on their ability to protect against cancer, especially prostate cancer.

Dr. Edward Giovannucci of the Harvard Medical School has published two interesting studies that investigated the effects of foods, particularly tomatoes, on cancer risk. In his 1995 study, Dr. Giovannuci found that of the 48,000 men surveyed, those who ate ten or more servings of tomatoes a week reduced their risk of prostate cancer by 35 percent and their risk of aggressive prostate tumors by almost 50 percent. Indeed, it seemed the higher the tomato intake, the lower the cancer risk. Interestingly, lycopene is the most abundant carotenoid in the prostate gland.

Dr. Giovannucci's subsequent study in 1999 showed that, of all tomato products, tomato sauce consumption—at just two servings a week—was by far the most reliable indicator of reduced risk for prostate cancer.

Two important points emerge from these studies. The first, which I mentioned earlier, is that processed tomatoes—sauce and paste—are more effective than raw tomatoes at reducing cancer risk. In the raw tomato, the lycopene is bound into the cell walls and fiber. Processing breaks down these cell walls and frees the lycopene to be absorbed by the body. Ounce for ounce, processed tomato products and cooked tomatoes contain two to eight times the available lycopene of raw tomatoes. While processing does diminish the levels of vitamin C in the tomatoes, it elevates the total antioxidant activity, thus ultimately providing an enhanced benefit.

• •

Tomato paste is a super ingredient to use in cooking. It has the power of a fresh tomato, but it is more concentrated! Eating tomato paste will increase the SPF (sun protection factor) of your skin, protecting you from damaging ultraviolet rays. In one study, ingesting 40 grams daily (less than one-quarter of a small can) of tomato paste, which provides about 16 milligrams of lycopene, resulted in a 40 percent increase in the amount of sun exposure it takes to cause reddening of the skin. Small amounts of tomato paste can enrich soups, stocks, and stews. Tomato paste comes in a can and squeezable tube. No-salt-added varieties are available. At our house we often double the amount of tomato paste called for in a recipe.

• •

The second important point, which Dr. Giovannucci mentions in his article, once again highlights the importance of whole foods. While he notes the association between tomato consumption and reduced cancer risk, particularly lung, stomach, and prostate cancers, he makes it clear that "a direct benefit of lycopene has not been proven and other compounds in tomatoes alone or interacting with lycopene may be important." Given the rich array of nutrients in tomatoes it wouldn't be surprising if, once again, the *synergy* of those nutrients were the reason for the positive effects.

Prostate cancer isn't the only type of cancer that tomatoes seem to help protect against. A growing body of evidence suggests that lycopene provides some degree of protection against cancers of the breast, digestive tract, cervix, bladder, and lung.

Lycopene seems to reduce the risk of cancer in several ways. As a particularly powerful antioxidant, it helps block the ongoing destructive effects of the free radicals in the body. It's especially effective in this mission when sufficient vitamin E is present. Lycopene also seems to interfere with the growth factors that stimulate cancer cells to grow and proliferate. And finally it seems to stimulate the body to mount a more effective immune defense against cancer.

As mentioned, lycopene, which is fat soluble, needs a bit of dietary fat to transport it into the bloodstream. A whole, fresh tomato, eaten out of hand, is not a good source of this nutrient. The top-ranked tomato-based foods that seem to be the most cancer-protective are all prepared with some oil. A salad of tomatoes with some extra virgin olive oil is really a health-promoting food. The green color of olive oil indicates the presence of polyphenols. Those polyphenols combined with the powerful nutrients in tomatoes are a healthy taste treat on spaghetti sauce, as a pizza topping, or in tomato-based soups.

••

Tomatoes All the Time

Compared to other carotenoids that are efficiently stored in the body, the plasma level of lycopene falls rather quickly when lycopene-rich foods are not included in the diet. Therefore, it seems prudent to get some lycopene from tomatoes in your diet on a daily basis if possible. Fortunately, this isn't difficult: most of us have an opportunity to eat a tomato-based food frequently. Think taco sauce, red marinara sauce, barbeque sauce, pizza, and even ketchup.

••

TOMATOES AND YOUR HEART

In addition to being cancer-protective, there's ample evidence that tomatoes also play a role in reducing your risk for cardiovascular disease. The antioxidant function of lycopene, combined with the other powerful antioxidants in tomatoes such as vitamin C and beta-carotene, work in the body to neutralize free radicals that could otherwise damage cells and cell membranes. This preservation of cells and their membranes reduces the potential for inflammation and thereby the progression and severity of atherosclerosis. In one study, German scientists compared the lycopene levels in the tissues of men who had suffered heart attacks with those of men who had not. The men who had suffered attacks had lower lycopene levels than those who hadn't.

Interestingly, the men with the lowest levels of lycopene were twice as likely to suffer a heart attack as those with the highest levels.

In another large European study, which compared carotenoid levels among patients from ten different countries, lycopene was found to be the most protective against heart attack.

Tomatoes are also a good source of potassium, niacin, vitamin B_6, and folate—a great heart-healthy combination of nutrients. Potassium-rich foods play a positive role in cardiovascular health, being especially effective in helping to achieve optimal blood pressure. Niacin is commonly used to lower elevated blood cholesterol levels. The combination of vitamin B_6 and folate effectively reduces levels of homocysteine in the blood. Elevated levels of homocysteine are associated with a higher risk of heart disease.

Pizza for the Prostate

The prostate gland does most of its growing between the ages of 13 and 20. If a boy eats lots of saturated fat during those years (the major "fast food" span for most Americans), it may increase his chances of developing prostrate cancer later in life. This is a strong argument for making pizza (double the sauce, halve the pepperoni, and sneak in some veggies) the fast food of choice for your kids, particularly your sons.

THE POWER OF SKIN

Growing out in the wild, even if it's the wild of your backyard, plants must protect themselves from attack. They're under constant assault from ultraviolet rays, pollution, and predators. It's important that they have a first, powerful line of defense. Skin is that defense. Whether it's the skin of an apple, the peel of a grape, or the rind on an orange, this part of the fruit has a tremendous antioxidant ability that permits it to withstand the assaults of nature. The outer leaves of spinach and cabbage, for example, have the highest levels of vitamin C, and broccoli florets have more C than the stalks. One hundred grams of fresh apples with the skin contain about 142 milligrams of flavonoids, but the same amount of apples without the skin has only 97 milligrams of flavonoids. Quercetin—a common flavonoid with anti-inflammatory properties—is found *only* in the skins of apples, not in the flesh of the fruit. The antioxidant activity of 100 grams of apples without the skin is 55 percent of the activity of 100 grams of apples with skin. The skinless apples are about half as powerful. The papery brownish skins on almonds and peanuts are loaded with various bioactive polyphenols.

As a general rule, the greater the proportion of skin to interior fruit, the higher the antioxidant ability. For example, blueberries and cranberries are extraordinarily high in antioxidants. The rule holds true for tomatoes: the smaller the tomato—think cherry tomato—the higher its antioxidant ability. You can use that antioxidant power by simply eating the skin! Try to eat appropriate fruits and vegetables with the skin on. Of course, the skin is where the pesticides and potentially harmful bacteria reside, so a careful washing is mandatory. Don't forget that juices with sediment on the bottom are the ones to choose. That sediment contains bits of skin and pulp and is a great source of antioxidants. You'll notice that many organic juices as well as those that are 100 percent juice contain this sediment.

Avoid cooking tomatoes in aluminum cookware. The acidity of the tomatoes may interact with the metal, causing it to migrate into your food, affecting the taste and possibly having negative effects on your health.

TOMATOES IN THE KITCHEN

Shopping for tomatoes that will have the most powerful, reliable effect on your health is surprisingly easy. As I've mentioned, processed tomatoes are actually of more benefit than fresh tomatoes. In fact, you probably already have in your pantry most of the tomato preparations that can help you boost your intake of this powerful food. Now, you simply have to remember to use them regularly!

Patients love it when I tell them that pizza can be a "health food." I always order it with extra sauce to boost my lycopene intake. It tastes great, too. If I'm eating pizza at home, I blot the slice with a paper napkin to soak up the fat.

Here are some quick ideas for getting tomato into your life:

- Sauté cherry tomatoes in some olive oil and herbs. Toss over pasta or serve as a side dish.
- Use sun-dried tomatoes (no salt added) in sandwiches.
- Toss a can of diced tomatoes into soups and stews.
- Make homemade pizza with extra sauce and top with your favorite vegetables. Many supermarkets sell pizza dough that only needs to be shaped, topped, and baked.
- A delicious quick meal is a turkey or chicken cutlet, pounded thin, quickly sautéed to brown it slightly,

topped with a favorite salsa, and baked in the oven until done. Sprinkle a bit of grated cheese on top near the end of cooking and shower it with chopped cilantro or parsley before serving if you like.

■ One of my favorite sandwiches is a toasted slice of whole wheat bread topped with sliced avocado and some chunky salsa.

ROASTED CHERRY TOMATOES

Cherry tomatoes
Extra virgin olive oil
Salt and pepper

Bake in the oven (450°F) for 20 minutes. Optional: sprinkle with fresh basil before serving.

Yellow and orange tomatoes do not contain lycopene. (Remember, lycopene is associated with the red pigment.) While these types of tomatoes do have other nutrients, like vitamin C, they are not good sources of lycopene.

See the SuperFoods Rx Shopping Lists (page 331) for some recommended tomato products.

Turkey (Skinless Breast)

SIDEKICKS: Skinless chicken breast
TRY TO EAT: 3 to 4 servings per week of 3 to 4 ounces (maximum, 4 ounces per serving)

••

Turkey contains:

- Low-fat protein
- Niacin
- Vitamin B_6
- Vitamin B_{12}

- Iron
- Selenium
- Zinc

••

At last! Turkey receives its just recognition. Passed over as the official national bird in favor of the eagle (despite Ben Franklin's enthusiastic support), the turkey is too often relegated to a once-a-year meal. Ignored and virtually invisible eleven months of the year, it's been a quiet few hundred years for the poor bird. Turkey is a SuperFood. Highly nutritious, low in fat, inexpensive, versatile, and always available, the turkey has finally come into its own. When you discover all of turkey's terrific nutritious benefits, it will surely become part of your regular diet.

Skinless turkey breast is one of, if not *the* leanest meat protein source on the planet. This alone could make it a SuperFood; but turkey also offers a rich array of nutrients, particularly niacin, selenium, vitamins B_6 and B_{12}, and zinc. These nutrients are heart-healthy and are also valuable in helping to lower the risk for cancer.

LOW-FAT PROTEIN

Because skinless turkey breast is so low in saturated fat, it closely approximates the lean sources of animal protein present during Paleolithic times. Studies suggest that the Paleolithic diet was a healthy one. There is also a broad consensus that the traditional Mediterranean/Japanese/Okinawa diets which are also low in saturated fat, have multiple health-promoting qualities. In general, there are no scientifically validated healthy dietary patterns that are high in saturated fat. There's no doubt that the leaner the protein source the better, but low-fat, healthy animal protein is very hard to find. Much of the poultry and red meat available in our markets has too much bad fat and little or no good fat. For example, 3 ounces of fresh ham has 5.5 grams of saturated fat. Three ounces of flank steak has 4.5 grams of saturated fat. The same amount of skinless turkey breast meat has less than 0.2 gram of saturated fat.

What about chicken? Many people think of chicken and turkey as virtually interchangeable. They're surprised when white-meat chicken breast isn't listed along with white-meat turkey breast as a SuperFood. But skinless roasted white meat of chicken is higher in calories and saturated fat than roasted white meat of turkey.

Meat (3 ounces)	Calories	Protein	Cholesterol*	Saturated Fat
Turkey (skinless, white meat)	115	26 grams	71 mg.	0.2 gram
Chicken (skinless, white meat)	140	26 grams	72 mg.	0.85 gram
95% lean ground beef	145	22 grams	65 mg.	2.4 grams

*Saturated-fat content is more important than the amount of cholesterol. You are "allowed" 300 mg. of cholesterol per day, and the amount of cholesterol in these meats is not high; by comparison one egg yolk has about 213 mg. of cholesterol.

PROTEIN IN OUR DIETS

Protein has become a loaded word these days. High protein, low carb . . . they're concepts that have dominated nutrition discussions in recent years. What's the truth? First, a very brief chemistry lesson: much of our body, including muscles, organs, skin, hair, and enzymes, is made primarily of protein. Protein is in every cell and is necessary for life. Protein, in turn, is composed of amino acids. Some amino acids are manufactured by the body. Nine others called essential amino acids must come from the foods we eat. Some foods, including all animal proteins, like eggs, meat, and fish, contain all of the essential amino acids and they're known as "complete" proteins. Other foods, particularly plant foods, are incomplete proteins; they must be made complete by getting their missing amino acids from other sources. That's why vegetarians must rely on certain combinations of foods, e.g., brown rice and beans, peanut butter and whole grain bread, and whole grain macaroni and cheese, in order to get complete protein. The only plant exception to this is soybeans and/or soy foods like tofu: they are complete proteins.

••

Don't worry too much about your total protein intake. Instead, think about healthy sources of protein and how to increase them in your diet. Remember that vegetarian sources of protein like soy foods and nuts and grains are good choices, as is seafood like salmon, oysters, clams, and sardines, and low-fat or nonfat dairy products.

••

Our bodies need a constant supply of protein. We don't store it as we do fats. However, getting enough protein isn't a problem for the vast majority of people. Most of us, in fact, get too much protein in our diets, or at least more than we need. The average woman eats 65 grams of protein daily; the average man eats 90 grams a day. Some high-protein diets recommend double or even triple that amount.

In 2002, the National Academy of Sciences published a new Dietary Reference Intake on everything from fiber to fatty acids. They recommend that, in order to reduce the risk of developing chronic degenerative diseases, an optimal range of protein intake is 10 to 35 percent of calories. (On a 2,000-calorie-a-day diet, this would be 50 to 175 grams of protein.) This recommendation is based on extensive scientific review and in my opinion it is an excellent guideline.

What does this translate to in everyday terms? Well, adult women need at least 46 grams of protein; adult men, 56 grams (very active and elderly people may well need more). It's very easy to achieve this protein recommendation. A woman can reach her daily goal with 3 ounces of tuna (20 grams of protein) plus 3 ounces of turkey breast (26 grams of protein). A slice of whole wheat bread has about 3 grams of protein and an ounce of almonds, 6 grams. Since many foods contain protein (a cup of lentil soup has about 7.8 grams; an egg, 6 grams; and a baked potato, 3 grams), you can see how quickly most people would reach their protein goal on a daily basis.

What about high protein diets? Many people mistakenly believe that there's some special "fat burning" paradise that you enter when you severely restrict your carbohydrate intake and simultaneously boost your protein intake. There is nothing magical about a high-protein diet, despite our eagerness to believe so. The simple, irrefutable fact is that if you eat more calories than you burn, you'll gain weight; if you burn more than you eat, you'll lose. Most people who follow a high-protein diet and lose weight do so simply because their food choices are such that they automatically cut down on calories. When you restrict or severely limit one group of foods (carbohydrates), a group that ordinarily comprises over half your calorie intake, you can't help but lose weight. And once you go off the diet, all or most of the weight usually comes back.

• •

Try to substitute nut and soy protein for some of your red meat consumption. This will definitely lower your risk for cardiovascular disease and possibly lower your risk for cancer.

• •

There are a few proven dangers in an exceptionally high protein intake. For one thing, the more protein you take in, the more calcium you excrete in your urine, thus raising your risk for osteoporosis. In the Nurses' Health Study, women consuming more than 95 grams of protein a day (an extra-lean 6-ounce hamburger has 48.6 grams of protein) had an increased risk of fractures. While there is ongoing debate on this subject, it seems that vegetable protein causes less bone loss than animal protein.

A high-protein diet is also associated with some risk for

kidney damage among susceptible people. If you have below-normal renal function, you should talk to your health care professional prior to trying a high protein diet.

Another danger of excessive protein in the diet has to do with insulin levels. One of the arguments for a high-protein diet is a claim that too many carbs raise the blood insulin level, which in turn causes weight gain by forcing the calories into fat cells rather than allowing these same calories to be burned as energy. A recent Michigan State University study seems to disprove this argument. In reality, higher insulin levels are a risk factor for developing diabetes and perhaps cancer.

••

Vegetarian Protein

It's relatively easy to consume sufficient protein if you're a vegetarian. If you select from two or more of these three groups in a given day, you're set:

Whole grains
Legumes
Nuts and seeds
••

Unfortunately, for most people in the United States, a high-protein diet means an increase in their consumption of red meat. It's this type of protein—with its associated saturated fat—along with the increased and disproportionate amount of it, that has the greatest negative impact on long-term health.

There is wide consensus that it is prudent to keep one's intake of saturated fat less than 7 percent of fat calories. Two significant sources of saturated fat in the typical American diet are red meat and full-fat dairy products. Numerous studies suggest there is a relationship between increased dietary saturated fat and colon cancer, coronary heart disease, and Alzheimer's disease. In addition, a number of studies have shown a link between red meat consumption and prostate cancer. It is also important to remember that saturated fat intake has a much stronger influence on increasing serum cholesterol than does dietary cholesterol intake. Substituting skinless turkey breast for higher-saturated-fat protein choices is an easy strategy to help lengthen your health span.

••

One complete protein source a day is enough. Scientists once believed that some complete protein was needed at each meal. We now know that the amino acids from protein remain in our bodies for at least four hours and for as long as forty-eight hours. So, don't worry about trying to eat complete protein at each meal. Think in terms of your daily intake.

••

WHERE'S THE BEEF?

There's nothing intrinsically wrong with red meat. Red meat from American buffalo, for example, is high in protein and low in saturated fat. The problem with most of the commonly available red meat in the United States is that it supplies too much of the fat we don't need—saturated fat and omega-6 fatty acids—and too little of the fat we do need—omega-3 fatty acids.

In theory, free-range and free-roam cattle should offer a better alternative to meat raised in feedlots. Cattle are ruminants, which means their digestive systems are primed for grass, not grain, but it's faster and easier to fatten them up on corn. A corn diet is rich in omega-6 fatty acids (which we get too much of), and so the meat of corn-fed cattle becomes high in this fatty acid, too. Corn-fed cattle will also tend to have higher residues of hormones and antibiotics.

••

Antibiotics are often used as growth promoters in farm animals. It is estimated that farmers administer approximately more than 26 million pounds of antibiotics to animals each year, with only about 2 million pounds used for treatment of infections. Remember, antibiotic-resistant strains of bacteria have been found in commercial meat products and in consumers' intestines! Let's lobby to lower the antibiotic use in animals that we use for food.

••

Grass-fed beef is leaner and has a healthier balance of omega-6 to omega-3 fatty acids. Grass-fed beef contains the plant-derived omega-3 fats and vitamin E found in green leafy vegetables, and it is lower in saturated fat compared with meat from corn-fed cattle. Of course, grass-fed, free-range beef is harder to find and more expensive than corn-fed supermarket beef. That's why leaner sources of protein—like turkey breast—make a better choice.

If you do choose beef of any type, trim all visible fat. On those occasions when my wife and I have ground beef, we rinse the browned beef in a strainer under hot water to get most of the fat out before adding spices and completing the recipe.

••

In the Mediterranean diet, red meat and meat products are consumed four to five times a month. Less is better, but this is a reasonable beginning goal for the majority of Americans. Ideally, you should have no more than 3 ounces of lean beef about every ten days.

••

TURKEY FOR YOUR HEART

Turkey is a good source of niacin, vitamin B_6, and vitamin B_{12}. All three of these B vitamins are important for energy production. Niacin seems to be associated with lowering the risk for heart attack and heart attack deaths. Low levels of vitamin B_{12}, as well as of B_6 and folate, are associated with high levels of homocysteine—an amino acid–like substance that may be an independent risk factor for heart disease.

TURKEY FOR YOUR IMMUNE SYSTEM

Turkey is rich in zinc, a remarkable nutrient that is present in all tissues of the body. Suboptimal zinc intake is common, and frequent consumption of turkey could play a role in improving overall zinc levels in the population. The zinc in turkey is far more bioavailable than the zinc in nonmeat sources of this important mineral. Zinc is critical for a healthy immune system. It helps promote wound healing and also normal cell division. A 3-ounce serving of turkey breast provides about 14 percent of your daily requirement for zinc.

TURKEY: SUPER SOURCE OF SELENIUM

Turkey is a good source of the trace mineral selenium. Selenium is of critical importance to human health. It is involved in a number of bodily functions, including thyroid hormone metabolism, antioxidant defense systems, and immune function. Evidence seems to suggest that there is a strong inverse relationship between selenium intake and the risk for cancer. The speculation is that this is due to selenium's role in DNA repair. Finally, there is epidemiological evidence that selenium intake may be related to a reduced risk of coronary artery disease. In areas of the country with a high level of selenium in the soil, there seems to be a lower incidence of this disease.

TURKEY IN THE KITCHEN

For the weeks leading up to Thanksgiving, food writers discuss turkey at a feverish pitch: frozen or fresh? How long to defrost? To brine or not to brine? And then turkey is forgotten until the next year. But things are changing, as turkey is beginning to increase in popularity. You do see recipes popping up that make use of this SuperFood. Years ago, turkeys were available primarily whole, but today you can buy turkey as whole or half breasts, cutlets, ground turkey, drumsticks, thighs, wings, and even tenderloins. All these parts cook quickly and make it easy to enjoy turkey frequently.

••

Fresh-ground turkey can be an excellent food, but read the label carefully. It should be ground from breast meat with no added skin. Sometimes it's ground including the skin, fat, and dark meat. Look for ground turkey that's 99 percent fat-free. If this figure is higher, it probably contains dark meat and/or skin.

••

Some of our favorite ways to use turkey:

- A turkey dinner made with a roasted whole fresh turkey breast. We remove the skin after cooking. It cooks quickly and goes with all the traditional trimmings.
- A turkey sandwich on toasted whole grain bread with spinach leaves and romaine lettuce, sliced onion and avocado, and a smear of mayo and/or mustard

- Turkey tacos or burritos using cooked, shredded turkey, stir-fried in olive oil with some onions and peppers
- Turkey slices with a bit of barbeque sauce (love that lycopene!). I take this to work in a Tupperware container with some cranberry sauce.
- Turkey soup with plenty of vegetables
- Lean, ground turkey breast meat in spaghetti sauce

••

Buy only whole turkey without added fats or oils. Self-basting birds may contain partially hydrogenated soy or corn oil or butter. Check the label carefully.

••

Walnuts

People have a predictable response when I tell them nuts are a SuperFood. Most say, "I can't eat nuts: they're too fattening." Some of my patients have said, "I can't even have nuts in the house. If they are around, I eat them." These responses are understandable; nuts are just plain delicious. My brother-in-law scarfed down as many nuts as possible after I told him about their considerable health benefits, and gained five pounds in one month. He tuned out when I told him the other part of the nut equation: moderation! Certainly nuts are high in calories, but they have extraordinary health benefits and are an important addition to your diet. And I'll give you some tips on how to enjoy them judiciously so you won't get fat.

First a simple fact: if you are overweight, smoke, never get off your sofa, and eat five fast-food meals a week, there's one thing you could do to improve your health and reduce your risk of cardiovascular disease without even taking your right hand off the remote. Eat a handful of nuts about five times a week. This simple act would reduce your chances of getting a heart attack by at least 15 percent and possibly as much as 51 percent. That's how powerful nuts are.

Nuts have attracted a great deal of attention lately. As a new nutritional era emerges that moves well beyond macronutrients like fat and protein and into the exciting world of phytonutrients, nutritionists are rediscovering these little nutrition powerhouses. I can safely say that nuts will play an important role in maximizing the human health span during this century.

It's a simple if astounding fact: people who eat nuts regularly can enjoy a significant reduction in their risk of developing coronary heart disease. They'll also reduce their risk of diabetes, cancer, and a host of other chronic ailments. One study that looked at all causes of death across various racial, gender, and age groups found an inverse ratio between nut consumption and *all causes* of mortality. If I could develop and patent a drug that could safely deliver the benefits of a daily handful of nuts, I'd be a billionaire!

You'll notice that walnuts are the flagship nuts in this *SuperFoods Rx* category. I want to stress, however, that all nuts and seeds are significant contributors to your good health. It makes sense that nuts and seeds are rich sources of a wide variety of nutrients. They are, after all, nature's nurseries. A nut or seed is basically a storage device that contains all the highly concentrated proteins, calories, and nutrients that a plant embryo will require to flourish.

While walnuts are the flagship *SuperFoods Rx* nut, my other two top choices include almonds and pistachios. My top-choice seeds are pumpkin and sunflower seeds.

Walnuts are the headliner for this category of SuperFood for a number of reasons. They are one of the few rich sources of plant-derived omega-3 fatty acids (called alpha linolenic acid, or ALA) along with canola oil, ground flaxseed and flaxseed oil, soybeans and soybean oil, wheat germ, spinach, and purslane. They are rich in plant sterols—plant sterols can play a significant role in lowering serum cholesterol levels—a good source of fiber and protein, and they also provide magnesium, copper, folate, and vitamin E. Finally, they're the nut with the highest overall antioxidant activity.

Peanuts are American's favorite nuts, even though they're not really nuts at all, but legumes closely related to beans. We consider them nuts here because they share a similar nutritional profile. Peanuts comprise two-thirds of our total nut consumption and rank third in snack-food sales in the United States. One ounce—about 48 peanuts—provides 15 percent of your daily vitamin E requirement, 2.5 grams of fiber plus calcium, copper, iron, magnesium, niacin, folate, and zinc along with 7 grams of protein.

Almonds are the best nut source of vitamin E, and a powerful plant source of protein. In fact, at 20 percent protein, one-quarter cup of almonds contains 7.6 grams of protein— more than a large egg, which contains 6 grams. Almonds also contain riboflavin, iron, potassium, and magnesium, and they're a good source of fiber. Almonds are also an excellent source of biotin, a B vitamin essential to the metabolism of both sugar and fat. One-quarter cup of almonds provides 75 percent of your body's daily requirement of this nutrient, which promotes skin health as well as energy levels. Almonds are also rich in arginine; only peanuts contain more. Due to its ability to promote the production of a specific chemical, arginine is a natural vasodilator, which promotes increased blood flow by relaxing the blood vessel wall. Also the skin of almonds has a number of polyphenols, many of which have significant free-radical scavenging properties.

· ·

Consumer Action Alert

Petition the large peanut producers to market raw or dry roasted peanuts *with the skins on* so we can get the considerable benefits of all those polyphenols.

· ·

Finally, almonds and peanuts also contain sphingolipids. At the moment, there's no known nutritional requirement for these lipids, but they seem to play an important role in cell membrane structure and function. Cancer involves numerous defects in cell regulation, and sphingolipids have been found to affect almost every aspect of defective-cell regulation in cancer. We need more information on this, but for now I believe it is safe to assume that sphingolipids play a role in optimizing our health and represent another component of whole foods that works in synergy with other nutrients and phytonutrients.

Pistachios are one of the oldest edible nuts on earth. In China, they are known as the "happy nut" because of their characteristic half-opened shell. A 1-ounce serving of pistachios equals 47 nuts—more nuts per serving than any other except peanuts at 48 per serving. Pistachios are loaded with fiber: you get more dietary fiber from a serving of pistachios than from a half-cup of broccoli or spinach. Pistachios are also rich in potassium, thiamine, and vitamin B_6. It's interesting to note that the B_6 in a 1-ounce serving of pistachios is equal to the B_6 in a typical 3-ounce serving of chicken or pork. Like all nuts, pistachios are particularly rich in the phytonutrients that are associated with reducing cholesterol and protecting from a variety of cancers.

••

Nuts in a Nutshell

A serving of shelled nuts is 1 ounce. One ounce of nuts is 10 to 48 nuts, depending on their size. A single serving of nuts provides between 150 and 200 calories.

••

SEEDS

While walnuts are the flagship food in this category, as we know by now, it's the synergy of multiple nutrients that provides the most benefits. Though seeds don't, as yet, have the research backing for their health benefits that nuts do, we can assume given their nutrient profiles that they also offer multiple health benefits. Even though we eat seeds in small quantities, they are a good source of protein, especially for vegetarians.

It's really a misnomer to separate nuts from seeds. In fact, just about any seed or fruit that contains an edible kernel inside a brittle covering is called a nut. So nuts and seeds include everything from almonds to walnuts as well as sunflower seeds, sesame seeds, pumpkin seeds, and pine nuts. For the purposes of *SuperFoods Rx*, my top picks include sunflower seeds and pumpkin seeds. I encourage people to include seeds in this food category and use them regularly as snacks, and in salads, cereals, and casseroles. I eat a handful of nuts and/or seeds every morning. I also prefer jams with seeds and I even eat watermelon seeds.

••

Super Sunflower Seeds

One ounce contains:
- 95 percent of the RDA for vitamin E
- Over 50 percent of the RDA for thiamine
- Close to 30 percent of the RDA for selenium
- 25 percent of the RDA for magnesium
- 16 percent of the RDA for folate

Sunflower seeds are:
- very high in polyunsaturated fatty acids
- a fairly rich source of potassium

••

Powerful Pumpkin Seeds

One ounce contains:

- Over 50 percent of the RDA for iron
- Over 30 percent of the RDA for magnesium
- Over 20 percent of the RDA for vitamin E
- About 20 percent of the RDA for zinc
- A generous amount of potassium

DON'T NUTS MAKE YOU FAT?

My brother-in-law will be the first to tell you that nuts do make you fat. Any food that you *add* to your current diet can make you fat. And it's true that nuts are high in calories. But the key concept with nuts is *substitution*. Add a few nuts to your daily diet, substituting them for other foods. You won't gain an ounce if you add 1 ounce of nuts at least five times a week and subtract a food of comparable calories, preferably one containing saturated fat like cheese or butter or, better yet, add the equivalent amount of calorie-burning exercise.

The truth is, those who eat nuts in a balanced diet tend to be thinner than those who don't, because nuts are so filling. Because of this, nuts help people stick to a diet of foods that are high in carbohydrates but low in fiber. In one Harvard study, people who ate 35 percent of their calories from healthy fats (the common recommendation is 25 to 30 percent of calories from fat) were three times more likely to maintain their weight loss than dieters who restricted their fat intake to 20 percent. While it's true that 79 percent of the energy of nuts comes from fat, it's also true that nuts are low in saturated fat and high in unsaturated fatty acids. Interestingly, saturated fat raises blood cholesterol levels about twice as much as polyunsaturated fats lower them.

Here's a breakdown of how many calories there are in common nuts followed by a few activities you can incorporate into your life to help you keep your nut intake in perspective. I have two recommendations on how to eat nuts and stay thin: substitute exercise for equivalent nut calories consumed or eliminate equivalent "bad" calories, say from saturated fat, from your diet.

Nut Calories
(All, except where noted, are for 1 ounce)

Almonds (24 nuts, raw)	164 calories
Almonds (22 nuts, dry roasted)	169 calories
English walnuts (14 halves)	185 calories
Hazelnuts (20 nuts, raw)	178 calories
Peanuts (48, dry roasted, no added salt)	166 calories
Peanut butter (2 tablespoons)	190 calories
Pecans (20 halves, raw)	195 calories
Pistachios (dry roasted, no added salt, 47 kernels)	162 calories
Pistachios (raw, no added salt, 47 kernels)	158 calories

Calories burned per activity
(Each activity would burn off approximately 150 calories, or one serving of nuts)

Walking briskly (4 mph)	32 minutes
Walking slowly	43 minutes
Running (6 mph)	13 minutes
Swimming (general)	21 minutes
Swimming (vigorous laps)	13 minutes
Cycling (vigorous, 14–16 mph)	13 minutes
Cycling (leisurely, 10–12 mph)	21 minutes
Stationary cycling (low setting)	26 minutes
Tennis, singles	16 minutes
Golf with cart	37 minutes

Golf, no cart, carrying clubs	16 minutes
Basketball	16 minutes
General gardening	26 minutes
Raking leaves	32 minutes

HOW NUTS HELP YOUR HEART

A powerful body of evidence now conclusively demonstrates that nut consumption correlates with reduced coronary artery disease. To date, at least five large epidemiological studies have demonstrated that frequent consumption of nuts decreases the risk of coronary artery disease. In each of these studies, the more nuts consumed (about five servings a week), the lower the risk. Even when adjusted for other factors, such as age, sex, race, and lifestyle variables, the results held. Overall, people who eat nuts five or more times a week had a 15 to 51 percent reduction in coronary heart disease. And amazingly, even people who ate nuts just once a month had some reduction.

One of the main contributors to heart health in nuts, particularly in walnuts, is the omega-3 fatty acids. We know that this particular component of fat works in various ways to help guarantee a healthy heart and circulatory system. Like aspirin, omega-3s "thin" the blood, helping it to flow freely and preventing clots from forming and adhering to the vessel walls. Omega-3s also act as an anti-inflammatory, preventing the blood vessels from becoming inflamed—a condition that reduces blood flow. Lowered blood pressure is another benefit of the omega-3s and, of course, reducing hypertension (or high blood pressure) is an excellent way to decrease your risk of cardiovascular disease and even macular degeneration.

Walnuts are also rich in arginine, which is an essential amino acid. Arginine helps to keep the inside of the blood vessels smooth while it also promotes the flexibility of the vessels, thus increasing blood flow, reducing blood pressure, and thereby alleviating hypertension. The top nut and seed sources of arginine in descending order include: watermelon seeds, pumpkin seeds, peanuts, almonds, sunflower seeds, walnuts, hazelnuts, and pistachios.

It's interesting to note that while the beneficial fatty acid composition of nuts would account for some of their positive effects on blood lipids, and thus their benefits to heart health, that doesn't explain the whole picture. In other words, in addition to the known health-promoting factors in nuts, including the omega-3s, the B vitamins, magnesium, polyphenols, potassium, and vitamin E, there are other elements, which we've yet to identify, that work to lower cholesterol levels and promote heart health.

A study of U.S. male physicians found an inverse association between nut consumption and sudden cardiac death. Doctors who consumed nuts were less likely to die from the sudden arrhythmias that often accompany a heart attack.

NUTS AND DIABETES

There was great excitement in the press very recently when researchers from Harvard studied more than 83,000 women and found that those who reported eating a handful of nuts or two tablespoons of peanut butter at least five times a week were more than 20 percent less likely to develop adult onset (type II) diabetes than those who rarely or never ate nuts. Type II diabetes develops when the body cannot properly use insulin. The women had been followed for up to sixteen years. The speculation is that the results apply to men as well as women. It's not only the "good" fat in the nuts that work on heart health. The fiber and magnesium in nuts help maintain balanced insulin and glucose levels. Certainly there couldn't be anything easier to do to improve your health than to eat a small handful of nuts five times a week!

••
Peanut Butter

Peanut butter, eaten in moderation, has good health benefits. Buy peanut butter from a manufacturer with good quality control; sometimes those grind-it-yourself arrangements that you may see in health food stores or farmers' markets, for example, are not as rigorously maintained in terms of cleanliness as they should be. Look for peanut butter with no added sugar, no salt, if possible, and, most important, with no partially hydrogenated oils. I like Laura Scudder's all natural old-fashioned peanut butter no salt added. I store it upside down in the pantry for a few days before I open it so the oil is dispersed throughout and I don't have to stir it quite so much.
••

NUTS TO THE RESCUE

While the evidence supporting nuts' contribution to heart health and diabetes prevention is impressive, we must remember that nuts, like every other SuperFood, don't just target a few isolated systems in our bodies. Indeed, they're categorized as SuperFoods because of their amazingly powerful effect on our overall health. Nuts are an extraordinary SuperFood because the range of their known health-promoting abilities is so wide. And surely there's even more to discover about the synergistic power of nuts.

Fiber: Nuts are a rich source of dietary fiber. In one study, a 10-gram-a day increase in dietary fiber resulted in a 19 percent decrease in coronary heart disease risk. One ounce of peanuts or mixed nuts provides about 2½ grams of fiber—a good contribution to overall daily fiber consumption.

Vitamin E: Most of us don't get nearly enough vitamin E in our daily diets, and nuts and seeds are a rich source of this nutrient. One of the components of vitamin E—gamma tocopherol—has powerful anti-inflammatory properties. It's my guess that this is one reason why nuts contribute so significantly to heart health, although studies still need to prove this. Different nuts provide different ratios of the components of vitamin E, which is a good reason to vary your nut consumption. Almonds, for example, are a rich source of alpha tocopherol. Pistachios, on the other hand, are low in alpha tocopherols but very high in gamma tocopherol and also contain gamma tocotrienols. Most vitamin E supplements, by the way, contain only d-alpha tocopherol and thus provide only a fraction of the benefit of vitamin E from whole food.

••

Nuts Go to Your Head

In 2002, the *Journal of the American Medical Association* published a report that found that a high dietary (not from supplements) intake of vitamins C and E may lower risk of Alzheimer's disease. In another study, vitamin E consumption was linked to a 70 percent reduction in the risk of developing Alzheimer's over a four-year period. Nuts are one of the richest dietary sources of vitamin E.

••

Folic Acid: This nutrient has gotten some attention lately because of its ability to prevent birth defects, particularly neural tube defects such as spina bifida. Nuts are rich in folic acid, whose benefits go beyond its critical role in birth-defect prevention. Folic acid also lowers homocysteine (an independent risk factor for cardiovascular disease) and helps prevent cancer and various causes of aging.

Copper: The copper in nuts is helpful in maintaining healthy levels of cholesterol. It also contributes to healthy blood pressure and helps prevent abnormal glucose metabolism.

••

Nuts to Wine

A handful of walnuts has significantly more polyphenols than a glass of apple juice or even a glass of red wine. In one study, a serving of shelled walnuts (fourteen halves) had the polyphenol content of 2.2 servings of red wine.

••

Magnesium: This important nutrient shows up in impressive amounts in nuts. Thus it's no surprise that low intakes of magnesium have been associated with increased risk of heart attack. Magnesium decreases heart arrhythmias and helps prevent hypertension. It's also critical for normal muscle relaxation, nerve impulse transmission, carbohydrate metabolism, and maintaining healthy tooth enamel. Low magnesium intake is also a risk factor for migraine headaches. It's interesting to note that almost half of patients who suffer from migraines have magnesium levels that are below normal.

Resveratrol: This flavonoid, which is found abundantly in grapeskins as well as peanut skins, has anticancer properties. It is also an anti-inflammatory and has been associated with helping to maintain healthy cholesterol levels.

Ellagic Acid: This polyphenol is found in high concentrations in nuts, particularly walnuts. Animal research studies have demonstrated that ellagic acid is beneficial in the prevention of cancer by affecting both the activation and detoxification of potential carcinogens.

NUTS IN THE KITCHEN

Because of their high concentrations of fats, nuts have a tendency to go rancid. Heat, humidity, and light all hasten their spoilage. Be sure you shop for nuts in a store with a high turnover, especially if you buy from bulk bins. All nuts should smell sweet or "nutty." A sharp or bitter smell indicates that the nuts may be rancid. In general, whole nuts keep better than pieces; unprocessed nuts keep better than processed ones; nuts in the shell keep better than shelled. Keep nuts in a cool place in a sealed container for up to four months. They'll keep in the fridge for about six months and in the freezer for up to a year.

Dry-roasted nuts are always a good choice. Avoid nuts with added oil. As always, check the labels: sometimes you'll find added preservatives, corn syrup, or other sweeteners plus salt. No-salt nuts are ideal, but if you can't eat nuts without salt, it's safe to enjoy them that way; just cut down on other sources of salt in your diet.

• •

You can roast nuts yourself but do so carefully: high temperatures destroy the omega-3s in the nuts. Spread the nuts on a cookie sheet and place in a 160 to 170°F oven for 15 to 20 minutes, or until they turn dark.

• •

Many of my patients have found that the freezer is the key to healthy nut consumption, especially if you're a nut nut who can't sleep if you think there's an open jar of mixed nuts in the house. (I use this trick with my daughter's oatmeal-raisin cookies; if they're out, they're eaten. If they're frozen, I have to give it a little thought before I eat one.) Try keeping a variety of nuts and seeds in heavy-duty freezer bags in the freezer. Then you can take out only small amounts at a time so that you'll have what you need—no more! If you prefer to keep them in the fridge, they should keep for about six months.

Most nuts taste better roasted or toasted. I often toss a handful of crushed walnuts or pine nuts or sliced almonds into a nonstick pan on medium heat while I'm assembling a salad. I shake the pan every few minutes until the nuts are lightly toasted, then toss them into the salad.

••

Here are some easy ways to make your life a little nuttier:

- Use nuts to top frozen yogurt.
- Toss some chopped nuts or seeds on top of a salad.
- Stir peanut butter into stews and curries to enrich and add flavor.
- Use finely chopped nuts to coat fish or poultry cutlets.
- Try peanut butter on pancakes with or without jelly.
- Gently sauté chopped nuts in olive oil along with bread crumbs and chopped garlic and toss with freshly cooked pasta.
- Don't forget that American classic—a peanut-butter-and-jelly sandwich. Make it on whole wheat bread; it's nutritious.
- Toss 2 tablespoons roasted sunflower seeds on your cereal.

••

Check the SuperFoods Rx Shopping Lists (pages 344–45) for some recommended nuts and seeds.

Yogurt

YOGURT SIDEKICKS: kefir
TRY TO EAT: 2 cups daily

••

Yogurt contains:

- Live active cultures
- Complete protein
- Calcium
- Vitamin B_2 (riboflavin)

- Vitamin B_{12}
- Potassium
- Magnesium
- Zinc

••

Remember the ads that featured those elderly people from the Caucasus mountain region of the Soviet Union who ascribed their extreme longevity to yogurt? Some had lied about their age to avoid conscription in the Soviet army. Many others simply realized that the older they claimed they were, the more excited their visitors became. Before you knew it, everyone in the neighborhood was nearly 120 years old, thanks to yogurt.

Those ads were created in a time when yogurt had to be "sold." It was assumed that no one would eat it if it didn't promise something remarkable. Times have changed. Today, we eat yogurt simply because we like it. But many of us have forgotten about the health benefits of yogurt, which were undiscovered or at least unproven in the days of those ads. And, because yogurt now comes in so many varieties and types—from frozen dessert bars to squeeze tubes of flavored yogurt—there are some facts that we need to know to reap the benefits of this extraordinary SuperFood.

THE SYNERGY OF PRE- AND PROBIOTICS

One of the most important aspects of yogurt as a source of health benefits is the synergy of two health-promoting substances it provides: prebiotics and probiotics.

Prebiotics are nondigestible food ingredients that beneficially affect the gut by selectively stimulating the growth and/or activity of one or more beneficial bacteria in the colon, thus improving host health. Fructooligosaccharides (FOS) are one of the many classes of prebiotics and they're found in legumes, vegetables, and cereals as well as yogurt. These nonabsorbed fibers inhibit potentially pathogenic organisms as well as increase the absorption of minerals such as calcium, magnesium, iron, and zinc.

Probiotics are defined as live microorganisms that, when taken in adequate amounts, can be of benefit to our health. The evidence for the role of prebiotics and probiotics in promoting health and fighting disease is increasing on a monthly basis and is now supported by many double-blind, placebo-controlled human trials. What used to be folklore has become scientific fact. This mounting body of very recent news simply confirms ancient wisdom. In 76 B.C., the Roman historian Plinius recommended fermented milk products (yogurt) for treatment of gastroenteritis. And a Persian version of the Old Testament (Genesis 18:8) states: "Abraham owed his longevity to the consumption of sour milk."

Flash forward to about a hundred years ago when Louis

Pasteur developed the germ theory of disease. He was one of the first to postulate that our health is intertwined with the living beneficial microorganisms residing on our skin and in our bodies. Yogurt is the most commonly eaten probiotic food that contributes to the balance of microorganisms in our system. With contemporary, cutting-edge research, folklore has become scientific fact: yogurt is indeed a SuperFood.

Like all of the SuperFoods, yogurt works synergistically to promote health and fight disease: it provides a range of health benefits that include live active cultures, protein, calcium, and B vitamins, which work together in such a way that the sum is greater than the parts. Yogurt's primary benefit—as a probiotic—is something that at first blush runs counter to the trend of most modern medicine. With the success of antibiotics beginning shortly after World War II, doctors and the public have come to view microorganisms as evil disease-promoters, which must be relentlessly eradicated. In fact, however, the key to health is *balance*: the goal is not to eradicate all microorganisms, but rather to promote the health of the beneficial ones. Yogurt plays a primary role in this promotion by encouraging the growth of "good" bacteria and limiting the proliferation of "bad" ones.

••

Yogurt has multiple immune stimulating activities both inside and outside the gastrointestinal (GI) tract. An interesting study has shown that if you eat yogurt with live active cultures, you decrease the amount of a common pathogenic bacterium— *Staphylococcus aureus*—in the nasal passages. This is a clear sign that the yogurt is stimulating the immune system, and there is a beneficial communication between the immune system lining the GI tract and the immune system lining the upper airway passages.

••

Our gastrointestinal tracts are home to over five hundred species of bacteria—some helpful and some harmful to our health. We rely on these beneficial microbial "partners" for a number of important functions, including carbohydrate metabolism, amino acid synthesis, vitamin K synthesis, and the processing of various nutrients. Yogurt is a source of beneficial bacteria, and the positive results that are ascribed to introducing this bacteria to our system are not relegated to the digestive tract. While a host of beneficial health effects are linked to yogurt, those that have attracted the most attention include its anticancer properties, its ability to lower cholesterol, and its ability to inhibit unfriendly bacteria.

One of the great benefits of the probiotics in yogurt is its ability to strengthen the immune system and thereby help the body prevent infection. In an era of antibiotic-resistant pathogens and seemingly new infectious threats like SARS and West Nile virus, the value of boosting one's immune system becomes immeasurable.

••

Inulin, a dietary fiber, is an additive used in Stonyfield Farm Yogurt. It's been shown to increase calcium absorption. For example, an intake of 8 grams of inulin a day increased calcium absorption among teenage girls by an average of 20 percent.

••

LIVE ACTIVE CULTURES

Before we explore yogurt's extraordinary abilities, it's important to understand that in order to be effectively health-promoting, the yogurt you buy *must contain live active cultures*. Yogurt is, quite simply, milk that has been curdled. To make yogurt, pasteurized, homogenized milk is inoculated with bacteria cultures and kept warm in an incubator where the lactose or milk sugar turns into lactic acid. This thickens the yogurt and gives it its characteristic tart, tangy flavor. The process is very similar to that used when making beer, wine, or cheese, in that beneficial organisms ferment and transform the basic food.

This is the basic process for producing yogurt, but there's a wide range of techniques adopted by manufacturers of differing brands. For example, some manufacturers pasteurize the yogurt after culturing it. In this case, the label will indicate "heat treated after culturing." This process kills all the friendly bacteria and, while it may taste good, its health benefits will not extend to those provided by live active cultures. You might be surprised to learn that some frozen yogurts have live active cultures. Check the labels; with live active cultures, frozen yogurt offers a low-fat advantage over ice cream.

• •

When you shop for yogurt, look for:

- Low-fat or nonfat varieties
- No artificial colors
- Very fresh (check the expiration date on the carton)
- Whey protein listed on the label (increases the viability of probiotic bacteria—as does the inulin in Stonyfield Farm yogurt)
- Rich in live active cultures (check for specific cultures; the more, the better)

• •

The National Yogurt Association has created a "live active cultures" (LAC) seal that guarantees that yogurt so labeled contain at least 100 million organisms per gram at the time of manufacture. "LAC" yogurt must be refrigerated and date-stamped to indicate its relatively short shelf life. After the expiration date on the yogurt, the bacteria numbers go down. Since the seal program is voluntary, some yogurt products may have live cultures but not carry the seal.

There are three basic types of yogurt, depending on the milk used to make it: regular yogurt, low-fat yogurt, and nonfat yogurt. Yogurt made from whole milk has at least 3.25 percent milk fat. Low-fat yogurt is made from low-fat milk or part-skim milk and has between 0.5 and 2 percent milk fat. Nonfat yogurt is made from skim milk and contains less than 0.5 percent milk fat. I favor nonfat yogurt.

••

Consumer Action Alert

Ask yogurt manufacturers to add vitamin D, a key player in calcium metabolism, to their yogurts.
••

One of the most common yogurts is the FOB, or "fruit on the bottom" product. Some FOB yogurts have live active cultures, but they also have a lot of added sugar. Some fruit-flavored yogurts have up to 7 teaspoons of sugar per cup! I don't like the taste of plain yogurt, so I usually get FOB yogurt. I leave the fruit on the bottom of the carton, but the yogurt still has some of the nice fruit taste. Ideally, the best yogurt to buy is plain nonfat or low-fat yogurt that is clearly labeled as containing "live active cultures." The label should also specify which cultures are in the product. The most popular yogurts use only two live cultures—*L. acidophilus* and *S. thermophilus*. Natural yogurts also include other beneficial bacteria, including *L. bulgaricus*, *B. bifidus, L. casei,* and *L. reuteri*. Be sure to check the label on your yogurt carefully—in general, the more beneficial cultures listed, the better. If you like fruit in your yogurt (and who doesn't?), add your own fresh or dried fruits to taste. I usually sprinkle mine with wheat germ, ground flaxseed, and berries for added taste and nutrition.

THE BENEFITS OF PROBIOTICS

Ultimately, it's yogurt's activity in the gastrointestinal tract that argues most conclusively for its inclusion as a Super-Food. The bottom line is that a healthy digestive system is critical to good health. Our ability to absorb nutrients from our food depends on our GI health. Even if we eat the most nutrient-dense foods in the world, if our digestive ability is impaired, we won't be able to benefit from those foods. As we age, our digestive ability is often diminished. All the more reason to rely on yogurt as a food that will promote and help preserve intestinal health.

PROBIOTICS AS DISEASE FIGHTERS

The list of the health-promoting abilities of probiotics is quite long. Some benefits have been proven absolutely conclusively while others require more study. Here is a summary of the conditions where yogurt has efficacy:

Cancer: Probiotics absorb mutagens that cause cancer, particularly colon cancer, though there's also evidence that they're effective in fighting breast cancer. They stimulate the immune system, partly by promoting immunoglobulin production, and help lower the risk for cancer by decreasing inflammation and inhibiting the growth of cancer-causing intestinal microflora.

Allergy: Probiotics are helpful in alleviating atopic eczema and milk allergy. In relation to eczema, it's important to remember that probiotics are working on promoting healthy skin as well as a healthy digestive tract. Indeed, probiotics affect all surfaces of the body that have interaction with the external world, including skin, nasal passages, gastrointestinal tract, and so forth. There is some evidence that babies who are exposed to probiotics (after the age of three months) will have a better chance of avoiding some allergies later in life.

Lactose Intolerance: Some people cannot tolerate milk because they lack the enzyme to break down milk sugar (lactose). In fact, only about a quarter of the world's adults can digest milk. This condition eliminates an important source of highly bioavailable calcium from the diet. Probiotics in yogurt digest the lactose for you, thus helping to relieve this condition. Yogurt is also a calcium- and vitamin-rich food that is readily digestible by those who suffer from lactose intolerance and is therefore an excellent addition to their diets.

Inflammatory Bowel Disease (IBD): Probiotics help regulate the body's inflammatory response, which relieves the symptoms of this condition. The probiotics in yogurt have been accepted as a form of therapy that can actually help maintain remission in people suffering from IBD. A 2003 review of human studies on probiotics, for example, concluded that "the use of probiotics in IBD clearly will not provide a panacea but it does offer hope as an adjunct form of therapy, specifically in maintaining a state of remission."

••

Older people, in particular, can benefit from yogurt. One research study tracked a population of 162 very elderly people for five years. Those who ate yogurt and milk more than three times per week were 38 percent less likely to die compared to those who ate those foods less than once a week. Yogurt helps people absorb nutrients, fight infection and inflammation, and get sufficient protein—all special challenges as we age.

••

Irritable Bowel Syndrome (IBS): Probiotics alter both the populations and the activities of the microflora in our gastrointestinal systems, possibly relieving the symptoms of IBS, though probiotics may prove to be more effective in prevention than in effecting a cure.

Hypertension: Probiotics stimulate the production of druglike substances that act in the body like pharmacological blood-pressure-lowering medicines.

Cholesterol Reduction: Over thirty years ago, scientists were intrigued to find that the Masai tribesmen of Africa had low serum levels of cholesterol as well as low levels of coronary heart disease, despite a diet that was extremely high in meat. The distinguishing characteristic of their diets, aside from high meat consumption, was an extremely high intake of fermented milk (or yogurt)—up to 5 liters daily. Research has now confirmed that yogurt is beneficial to those trying to reduce cholesterol. The probiotics in yogurt reduce the bile acids, which in turn decrease the absorption of cholesterol from the gastrointestinal tract. This effect seems to be seen most reliably in people who already have elevated cholesterol.

Ulcers: Probiotics help to eliminate the pathogen Helicobacter pylori, a bacterium that is one of the main causes of ulcers and may also be a cause of gastric cancer.

Diarrhea: Yogurt has potential benefit in relieving what in many countries around the world is a serious threat to the health of millions. It fights diarrhea by stimulating the immune system, crowding out negative microflora in the intestines and stimulating the growth of beneficial bacteria. Probiotics in yogurt are also helpful in treating diarrhea associated with antibiotic use, and some doctors are amazed that yogurt is not routinely recommended to all patients who are being treated with antibiotics.

••

Consumer Action Alert: We All Need More Culture!

Many yogurts claim to have live active cultures. Mainstream yogurts must contain *L. acidophilus* and *S. thermophilus* in order to be so labeled. Some yogurts add more, including *L. bulgaricus*, *B. bifidus*, *L. casei*, and *L. reuteri*. Look for yogurts that contain the most variety of live active cultures. Since certain species of probiotic cultures have health benefits, let's urge manufacturers to put them in commonly available yogurts.

••

Vaginal Infections and Urinary Tract Infections: Once again, the probiotics in yogurt fight pathogens while crowding out the "bad" microflora and stimulating the growth of beneficial bacteria. One study concluded that eating 8 ounces of yogurt containing *L. acidophilus* on a daily basis decreases candidal yeast colonization and infection threefold when compared with control groups.

YOGURT: BEST IN DAIRY CLASS

Most people are surprised to learn that in the United States, *nine out of ten women and seven out of ten men don't meet their daily requirement for calcium.* Worse and even more troubling news is that nearly 90 percent of teenage girls and 70 percent of teenage boys don't meet their daily calcium requirement. For many, soda has replaced the old "milk at every meal" custom. This portends disastrous future health consequences for large numbers of people. A single 1-cup serving of nonfat plain yogurt supplies 414 milligrams of calcium—an amazing 40 percent of your daily calcium needs and at a cost of only 100 calories. This compares favorably with nonfat milk, which has only 300 milligrams of calcium. The rich amount of potassium in yogurt combined with the calcium also plays a role in normalizing your blood pressure.

Yogurt is also a better source of B vitamins (including folate), phosphorus, and potassium than milk. Of course, the calcium in yogurt is of great benefit to pre- and post-menopausal women and to men in their struggle against osteoporosis. A rich source of calcium to begin with, the milk sugar in yogurt actually aids in calcium absorption. Moreover, dairy foods are a source of IGF-1, a growth factor that promotes bone formation, which benefits women over and above the bone-preserving contribution of calcium.

••

Best Quick All-Star SuperFoods Rx Breakfast

One of my breakfast favorites couldn't be easier to fix. I take a bowl of nonfat yogurt and top it with a handful of blueberries (and/or raspberries, cherries, or whatever fruit is in season) and some sliced banana. I toss in a small handful of chopped walnuts and about a tablespoon of wheat germ or ground flaxseed. It's delicious and nutritious.

••

YOGURT: GREAT SOURCE OF DIGESTIBLE PROTEIN

Yogurt is a great source of readily digestible protein. In fact, yogurt supplies double the protein of milk because it's usually thickened with nonfat milk solids, increasing its protein content. Some people, particularly the elderly, just don't consume enough protein or calcium. Studies have shown there's a positive association between protein intake and bone-mineral density of older women and men when they're supplemented with calcium. The lesson: optimum bone health and prevention of osteoporosis depend not just on calcium supplementation, but on sufficient protein intake as well. Yogurt, with its easily digestible protein and calcium, is the answer.

Make Yogurt Cheese

Line a sieve with a coffee filter and drain the yogurt for a few hours in the fridge—the longer it drains, the thicker it becomes. Use the resulting liquid or whey in pancakes or muffins as a milk substitute. Mix the yogurt cheese half and half with mayo in tuna or other salads to reduce fat and boost protein. Use yogurt cheese in dips, spooned onto a bowl of chili, or as a fruit topping.

BLUEBERRY YOGURT SHAKE

MAKES 2 SERVINGS

1 cup nonfat plain yogurt
¼ cup freshly squeezed orange juice
½ cup fresh of frozen blueberries
½ very ripe banana

Combine all ingredients in a blender. Blend on medium speed until smooth and frothy. Pour into glasses and serve.

See the SuperFoods Rx Shopping Lists (pages 353–54) for some recommended yogurt foods.

PART III

The SuperFoods Rx Menus and Nutritional Information

The SuperFoods Rx Menus

If you have ever visited Rancho La Puerta or the Golden Door, you know what a remarkable experience they provide. It's not just the utter tranquility and beauty of the locations; the combination of attention to the body with exercise and spa treatments, attention to the spirit with an overall atmosphere of mindful serenity, and delicious, healthful food refreshes one entirely. It's hard to imagine how a single week can echo for months and months, but it does. My recent week at Rancho La Puerta is such a strong memory that I can close my eyes in the most stressful moment and be transported back to the total peace of that week. I hope that every reader of this book will someday have the opportunity to enjoy a comparable experience.

When the idea for *SuperFoods Rx* first took shape, I realized that I would need more than just nutritional statistics to convince people to adopt the *SuperFoods Rx* lifestyle. Dr. Hugh Greenway, my friend and colleague, suggested that the world-renowned chef Michel Stroot, of the Golden Door, and his colleagues at Rancho La Puerta would be ideal candidates to create recipes for *SuperFoods Rx* that would be delicious and easy to prepare. Both spas are world famous for their delicious and healthful meals. Dr. Greenway has had a long relationship with both Rancho La Puerta and Golden Door because he treated Alex Szekely, son of the founders of both spas, when he was first diagnosed with cancer. Alex sadly lost his battle with melanoma, but his legacy lives on in both spas. He brought his parents' vision of a spa retreat to another level when he shifted the emphasis from weight loss and general pampering to retreats that advocate the ideal of a mind-body-spirit balance.

Rancho La Puerta and the Golden Door and their talented chefs, particularly Michel Stroot, immediately grasped the theory of *SuperFoods Rx*—that if you shifted the emphasis of your diet to health-promoting foods, you'd feel better, avoid many causes of illness and death, and you'd even look better as a delightful bonus.

The recipes included here are the best of their creations, working within the guidelines of the *SuperFoods Rx* program. I'm thrilled that they have helped my vision come to life and hope you and your families will enjoy these recipes.

••

For more information about the Golden Door, you can phone them at 760-744-5777 or visit their website at: www.goldendoor.com. For more information on Rancho La Puerta, you can phone them at 760-744-4222 or visit their website at: www.rancholapuerta.com.

••

DAY 1

MENU

Breakfast: Oatmeal, fruit, nuts, and soymilk
Morning Snack: 1 cup papaya chunks

Lunch: Baked Acorn Squash with Quinoa, Tofu, Dried Apricots, and Walnuts; Tomato and Cucumber Salad
Afternoon Snack: ½ ounce dry roasted or raw almonds, 8 ounces Knudsen Very Veggie Cocktail, 1 small carrot
Dinner: Grilled Filet Mignon with Cremini Mushrooms on Wilted Greens, Scalloped Sweet Potatoes, SuperFoods Rx Salad (page 273), Watermelon-Banana Sorbet

■ *Breakfast*

Cook ½ cup oatmeal as directed on the side of the box. Add 1 tablespoon ground flaxseed meal, 2 tablespoons toasted pecans, a handful of raspberries, and ½ cup soymilk.

Drink 1 cup of unsweetened grape juice.

■ *Lunch*

BAKED ACORN SQUASH WITH QUINOA, TOFU, DRIED APRICOTS, AND WALNUTS

SERVES 4

Apricot sauce
   ¼ cup dried apricots, cut in half
   1 cup water
   1 tablespoon frozen orange juice concentrate

Baked acorn squash
   2 medium acorn squash
   1 teaspoon onion powder

2 cups plus 2 tablespoons vegetable stock or water
1 teaspoon sea salt
1 cup quinoa, rinsed and drained
½ cup minced fresh flat-leaf parsley
½ cup coarsely chopped walnuts
½ pound firm tofu
1 teaspoon low-sodium soy sauce
½ teaspoon dried oregano
1 teaspoon olive oil
½ cup quartered dried apricots

1. To make the sauce, simmer the apricots and water over medium heat for 10 to 12 minutes, or until the liquid is reduced by half. Set aside to cool.

2. Transfer the cooled mixture to a blender, add the orange juice concentrate, and blend until smooth. Set aside until ready to use. Reheat gently, if necessary.

3. Preheat the oven to 375°F.

4. To prepare the squash, cut them in half horizontally and scoop out the seeds. Trim the bottom of the squash halves so that they sit level. Put in a baking dish and sprinkle with onion powder. Pour cold water in the dish to a depth of ½ inch. Cover with foil and bake for about 50 minutes, or until tender. Add more water if necessary during baking. Remove from the dish.

5. Meanwhile, in a saucepan, bring 2 cups of the vegetable stock and the salt to a boil over medium-high heat. Add the quinoa, cover, and simmer for about 20 minutes. Remove the pan from the heat and let stand, still covered, for about 10 minutes. Toss with the parsley and walnuts and fluff with a fork. Cover and set aside to keep warm.

6. Cut the tofu into ½-inch dice, put into a bowl, add the soy sauce and oregano, and toss gently.

7. In a nonstick sauté pan, heat the olive oil over

medium-high heat. Add the tofu and any liquid from the bowl and sauté for about 4 minutes. Add the dried apricots and the remaining 2 tablespoons of stock to the pan. Stir to scrape up any browned bits on the bottom of the pan.

8. Gently toss the tofu and apricots with the quinoa. Spoon into acorn squash and drizzle with the apricot sauce.

TOMATO AND CUCUMBER SALAD

SERVES 4

Balsamic vinaigrette
2 tablespoons balsamic vinegar
1 tablespoon olive oil
1½ tablespoons Dijon mustard
1½ tablespoons water
½ teaspoon dried basil
¼ teaspoon freshly ground black pepper

Salad
4 ripe medium tomatoes
3 medium cucumbers

1. In a small bowl, whisk the vinegar and oil. Add the mustard and water and whisk until blended. Whisk in the basil and pepper. Whisk again before using.

2. Core the tomatoes and cut into wedges. Cut each wedge in half. Peel the cucumbers and slice into ½-inch-thick rounds. Quarter the rounds.

3. Toss the tomatoes and cucumbers in a large bowl. Pour the dressing over them, stir gently to mix, and then set aside at room temperature to marinate for at least 10 minutes or up to 1 hour.

GRILLED FILET MIGNON WITH CREMINI MUSHROOMS ON WILTED GREENS

SERVES 4

1 cup stemmed, quartered cremini mushrooms
2 tablespoons minced shallots
½ cup red wine, such as Merlot
½ cup Beef Brown Sauce (page 249), or low-sodium canned beef gravy
2 tablespoons olive oil
½ cup finely chopped yellow onion
2 teaspoons minced garlic
4 cups torn turnip greens
1 pound beef tenderloin or filet, sliced into four 4-ounce pieces about 1-inch thick, or bison filets
2 tablespoons chopped fresh flat-leaf parsley

1. Spray a nonstick sauté pan with canola oil spray, set over medium heat, and sauté the mushrooms and shallots for about 5 minutes, or until lightly browned. Add the wine and cook for about 10 minutes, or until reduced by half. Add the brown sauce, bring to a simmer, and cook for 2 to 3 minutes, or until the flavors blend.

2. In another large sauté pan, heat the oil over medium-high heat and cook the onion and garlic, stirring, for 2 to 3 minutes, or until the onions become translucent. Add the turnip greens and cook, stirring, for a few minutes until the greens wilt. Remove from the heat and keep warm.

3. Prepare a charcoal or gas grill. The coals should be medium-hot to hot. Lightly spray the grill rack with vegetable oil spray.

4. Grill the filets for about 5 minutes on each side until they reach the desired degree of doneness. Insert an instant-read thermometer in a filet. After about 5 minutes, the temperature will be 145°F. for medium-rare; after 6 to 7 minutes, the temperature will be 160°F. for medium. Remove from the grill.

5. To serve, fork the wilted greens onto a plate and top with the grilled steak. Spoon the mushroom sauce over the meat. Garnish with chopped parsley.

BEEF BROWN SAUCE

MAKES ABOUT 1½ CUPS

2½ pounds top round, including trimmings, or bison, cut into 1-inch pieces
1 small onion, coarsely chopped
1 medium carrot, coarsely chopped
1 celery rib, coarsely chopped
1 tablespoon tomato paste
2 teaspoons crushed black peppercorns
2 teaspoons dried tarragon
2 sprigs fresh thyme or 1 teaspoon dried thyme
2 whole bay leaves
2 tablespoons unbleached all-purpose flour
2 quarts vegetable broth or water
1 tablespoon arrowroot or cornstarch, dissolved in 2 tablespoons water

1. Spray a large skillet with canola oil, set it on medium-high heat, and sauté the beef trimmings for 10 to 15 minutes, or until browned on all sides. Add the onion,

carrot, and celery and sauté for 5 to 10 minutes longer, or until well browned.

2. Stir in the tomato paste, peppercorns, tarragon, thyme, and bay leaves, and sprinkle with the flour. Transfer to a large pot.

3. Pour 2 cups of the vegetable broth into the skillet and scrape off any caramelized pieces of meat or vegetables. Pour into the pot with the vegetables. Add the remaining vegetable broth and simmer for about 2 hours, skimming the foam that rises to the top. Simmer until reduced by about two-thirds. Strain through a fine-mesh sieve into another pot.

4. Return to the stove, bring to a simmer, and cook for about 30 to 40 minutes, or until reduced to 1½ cups. Stir in the arrowroot or cornstarch mixture and simmer for another 1 to 2 minutes until thickened.

SCALLOPED SWEET POTATOES

SERVES 4

2 medium sweet potatoes (about 1 pound)
¼ cup fresh orange juice
¼ cup vegetable stock or water
1 teaspoon ground allspice

1. In a medium saucepan, combine the unpeeled sweet potatoes and enough cold water to cover and bring to a boil over high heat. Reduce the heat and simmer for 15 to 20 minutes, or until the potatoes just begin to soften but are still relatively firm. Drain and set aside until cool enough to handle.

2. Peel the sweet potatoes and cut into ½-inch-thick rounds. You will have about 16 slices. Place 4 sweet potato slices in a medium sauté pan, arranging them in a circle and overlapping the edges slightly. Repeat the pattern with the remaining slices. Drizzle the orange juice and stock over the potatoes and sprinkle with allspice. Cover and set aside until ready to serve. Reheat in the oven if necessary.

WATERMELON–BANANA SORBET

SERVES 4

2 large ripe bananas, peeled and thinly sliced
¾ cup watermelon cubes
½ cup freshly squeezed orange juice
⅓ cup fresh lime juice

1. Spread the bananas and watermelon in a shallow metal dish. Pour the orange juice and lime juice over the fruit and freeze for at least 4 hours, or until frozen solid.

2. Let the fruit thaw on the countertop for about 10 minutes. Break into small pieces and transfer to a food processor fitted with a metal blade and process until smooth and creamy.

3. Spoon into small cups and serve immediately or return to the freezer for no more than an hour. If you freeze the sorbet for longer than an hour, process it again in the food processor before serving.

MENU

Breakfast: "Fortified" cereal with soymilk, 8 ounces pink grapefruit juice
Morning Snack: 1 cup fresh papaya chunks
Lunch: SuperFoods Spinach Pasta with Turkey-Tomato Sauce, SuperFoods Rx Salad (page 273) with Raspberry Vinaigrette
Afternoon Snack: ½ yellow bell pepper, cut into strips
Dinner: Asian-Style Wild Salmon on Sesame Spinach, Millet and Steamed Asparagus Tips, Berry Crisp with Nuts and Oatmeal Topping

▪ *Breakfast*

Mix together ¾ cup Kashi Good Friends cereal, 1 tablespoon toasted wheat germ, 2 tablespoons toasted slivered almonds, 2 tablespoons wheat bran, and ½ tablespoon ground flaxseed meal. Pour 8 ounces vanilla soymilk over the cereal. Enjoy a 4-ounce glass of unsweetened pink grapefruit juice while mixing together your morning mélange.

SUPERFOODS SPINACH PASTA WITH TURKEY-TOMATO SAUCE

SERVES 6

9 medium ripe tomatoes (about 3 pounds), halved or quartered, or two 16-ounce cans whole tomatoes with juices, plus two 8-ounce cans tomato sauce

½ cup water

One 8-ounce can tomato paste

2 tablespoons olive oil

1 cup chopped onion

2 to 3 garlic cloves, minced

1 pound lean ground turkey

1 small fennel bulb, chopped (about 1 cup), or 1 cup chopped celery plus ½ teaspoon crushed fennel seed

2 tablespoons finely chopped fresh flat-leaf parsley or 2 teaspoons dried parsley

2 tablespoons chopped fresh basil or 2 teaspoons dried basil

1 tablespoon chopped fresh oregano or 1 teaspoon dried oregano

¼ teaspoon freshly ground black pepper

¼ to ½ teaspoon coarse sea salt

Pinch of cayenne

1 pound spinach pasta, freshly cooked

1. In a large nonstick pot, cook the tomatoes and water over low heat for 30 to 40 minutes, or until tender. Transfer to a blender or food processor fitted with a metal blade, add the tomato paste and process until smooth. (If using canned tomatoes, pour them and the canned tomato sauce and paste

directly into the blender or food processor.) Work in batches if necessary.

2. In a large sauté pan, heat the olive oil over medium-high heat. Add the onion and garlic and cook, stirring, for about 5 minutes, or until softened but not browned. Add the turkey and cook for about 10 minutes, breaking up the turkey with a spoon into small pieces. Add the fennel and cook for 1 to 2 minutes longer, or until the fennel softens. Pour contents of pan into strainer and allow fat to drain.

3. Return meat to pan (or heavy pot), add the tomato puree, parsley, basil, oregano, black pepper, salt, and cayenne. Cover and cook for about 45 minutes, or until flavors are nicely melded.

4. Serve over spinach pasta.

RASPBERRY VINAIGRETTE

½ cup raspberries
3 tablespoons balsamic vinegar
1 tablespoon unsweetened apple juice
1 teaspoon canola oil
1 teaspoon honey

Whisk the ingredients together in a small bowl. Toss the dressing with the salad.

ASIAN-STYLE WILD SALMON ON SESAME SPINACH

SERVES 6

Marinade

½ cup rice wine vinegar
½ cup water
6 tablespoons low-sodium soy sauce
2 teaspoons toasted sesame oil
Juice of 2 small limes
1 knob fresh ginger, about 1½ to 2 inches long, peeled and
    minced
Six 4-ounce wild salmon fillets

Sauce

1 tablespoon chopped shallots
1 teaspoon minced garlic
1 teaspoon minced ginger
¼ cup minced fresh cilantro leaves

Nests

½ teaspoon sunflower or safflower oil
1 tablespoon chopped shallots
1 garlic clove, minced
2 bunches spinach (each 10 to 12 ounces), washed,
    stemmed, and chopped
1 tablespoon sesame seeds

Garnishes

¼ cup finely sliced chives
¼ cup finely minced red bell pepper
Millet and Steamed Asparagus Tips (page 257)

1. To make the marinade, combine the vinegar, water, soy sauce, oil, and lime juice in a blender. Add the ginger and blend on high speed for 2 minutes, or until smooth.

2. Lay the salmon fillets in a shallow glass or ceramic dish and pour half the marinade over them. Cover and refrigerate for 30 minutes. Reserve the remaining marinade.

3. Preheat the oven to 350°F.

4. Lift the salmon from the marinade and discard the marinade. Lay the fillets in a shallow baking dish and bake for 15 to 20 minutes, or until the fillets just begin to flake when pierced with a fork and an instant-read thermometer registers 140°F.

5. Meanwhile, put the reserved marinade in a saucepan, add the shallots, garlic, and ginger, and bring to a simmer over medium heat. Cook for about 5 minutes, or until the shallots and garlic are tender. Stir in the cilantro, cook for a minute or so, and then remove from the heat.

6. To make the nests, in a large sauté pan, heat the oil over medium heat and cook the shallots and garlic until softened. Add the spinach and cook just until wilted. Sprinkle with sesame seeds.

7. Divide the spinach among 6 plates and top each mound of spinach with a salmon fillet. Spoon the sauce over the salmon and garnish with a sprinkling of chives and minced red peppers. Serve with millet and asparagus tips.

MILLET AND STEAMED ASPARAGUS TIPS

SERVES 6

    1 cup millet
    2 cups vegetable stock or water
    1 teaspoon kosher salt
    Eighteen 2-inch-long asparagus tips

1. Rinse the millet in a fine-mesh sieve under cool running water. Drain.

2. In a saucepan, bring the stock and salt to a boil over medium-high heat. Add the millet, stir, reduce the heat to a simmer, cover, and cook for about 20 minutes, or until the grain is tender. Remove from the heat and fluff with a fork.

3. Meanwhile, put the asparagus tips in a steaming basket and set over boiling water. Steam for 2 to 3 minutes, or until crisp-tender.

4. Serve the millet with the asparagus tips on the side or on top.

BERRY CRISP WITH NUTS AND OATMEAL TOPPING

SERVES 8

Crisp
    ⅓ cup chopped almonds
    ⅓ cup chopped pecans
    ⅓ cup chopped walnuts
    1 cup quick-cooking or regular rolled oats
    3 to 4 tablespoons pure maple syrup

2 tablespoons wheat germ
2 tablespoons whole wheat pastry flour or unbleached
   all-purpose flour
1 teaspoon cinnamon
½ teaspoon nutmeg
½ teaspoon pure vanilla extract

Filling

4 cups blackberries, blueberries, strawberries, raspberries,
   or a mixture of any of these
2 tablespoons pure maple syrup
1 teaspoon ground cinnamon
1 teaspoon finely grated lemon or lime zest
16 ounces nonfat regular or frozen vanilla yogurt, for
   topping

1. Preheat the oven to 325°F.

2. To prepare the crisp, spread the nuts on a baking sheet and toast for 5 to 8 minutes, or until golden brown. Stir once or twice during toasting. Transfer the nuts to a mixing bowl.

3. Add the oats, syrup, wheat germ, flour, cinnamon, nutmeg, and vanilla. Mix well.

4. For the filling, slice any of the berries that need it. Toss with the syrup, cinnamon, and zest. Spread the fruit in an 8- or 9-inch square baking pan or pie plate. Top with the crisp and bake for 15 to 20 minutes, or until the fruit is tender, bubbling around the edges, and the crisp is lightly browned. Serve topped with a 2-ounce dollop of yogurt.

MENU

Breakfast: Tropical Fruit Yogurt Parfait, 4 ounces orange juice, Whole Grain Toast with Almond Butter
Lunch: Poached Wild Salmon Salad with Cucumber Dill Sauce, Sweet Potato Oven Fries
Afternoon Snack: ½ orange bell pepper, cut into strips; ½ ounce unsalted, dry-roasted soy nuts; 1 small carrot
Dinner: Stir-Fried Tofu with Soba Noodles, SuperFoods Rx Salad (page 273), Golden Door Blueberry Bread

■ *Breakfast*

TROPICAL FRUIT YOGURT PARFAIT

SERVES 4

½ medium papaya, seeded, cut into ½-inch cubes (about 1 cup)
3 cups nonfat vanilla yogurt (or plain nonfat yogurt mixed with 2 tablespoons pure maple syrup)
4 tablespoons ground flaxseed
½ cup sliced almonds
1 cup blueberries
1 mango, cut into ½-inch cubes (about 1 cup)

1. Spoon about 2 tablespoons of the papaya cubes into each of four (12-ounce) parfait cups or tall, wide-mouth glasses. Top each with about 2 tablespoons yogurt, 1 teaspoon ground flaxseed, and 1 teaspoon almonds.

2. Spoon 2 tablespoons blueberries into each parfait cup and top each with about 2 tablespoons yogurt, 1 teaspoon ground flaxseed, and 1 teaspoon almonds. Spoon about 2 tablespoons mango cubes into each parfait cup and top each with about 2 tablespoons yogurt, 1 teaspoon ground flaxseed, and 1 teaspoon almonds. Repeat with remaining ingredients to make 3 more layers of fruit and yogurt.

3. Refrigerate or serve immediately with Whole Grain Toast with Almond Butter (below) and ½ cup fresh, full-pulp orange juice.

WHOLE GRAIN TOAST WITH ALMOND BUTTER

MAKES 4 SLICES

    4 slices whole grain bread
    1 tablespoon almond or cashew butter

Toast the bread slices. Spread each with almond butter, cut in half diagonally, and serve.

NOTE: Cashew or hazelnut butter or other tree nut butter may be substituted.

POACHED WILD SALMON SALAD WITH CUCUMBER DILL SAUCE

SERVES 6

Sauce

- 1 cup nonfat yogurt
- ½ cup low-fat cottage cheese, ricotta, or nonfat yogurt cheese (see page 240)
- ¼ cup chopped fresh flat-leaf parsley
- 2 tablespoons chopped fresh dill or 1 tablespoon dried dill
- 2 medium cucumbers, peeled, seeded, and shredded
- ¼ cup minced shallots or onion
- Pinch of cayenne
- ¼ teaspoon sea salt
- ¼ teaspoon freshly ground black pepper

Salmon

- 1 cup white wine
- ½ cup vegetable stock or water
- Juice of 1 lemon
- 1 shallot or small onion, minced
- 1½ teaspoons dried thyme
- ¼ teaspoon cracked black pepper
- Six 4-ounce wild salmon fillets

Assembly

- 6 cups washed and drained spinach leaves
- 2 teaspoons olive oil
- 2 teaspoons balsamic vinegar or fresh lemon juice
- 1 small unpeeled cucumber, thinly cut into rounds
- 6 large cherry tomatoes, for garnish

1. To prepare the sauce, combine the yogurt and cottage cheese in a blender or food processor fitted with a metal blade and process until smooth. Scrape into a bowl and stir in the parsley, dill, cucumber, shallots, cayenne, salt, and pepper. Taste and adjust the seasoning. Cover and refrigerate for at least 2 hours.

2. To prepare the salmon, combine the wine, stock, lemon juice, shallot, thyme, and black pepper in a large sauté pan and bring to a boil over medium-high heat. Reduce the heat to low and simmer the poaching liquid for 2 to 3 minutes. Add the salmon and poach for 13 to 15 minutes, or until the fish flakes easily when pierced with a fork, and an instant-read thermometer registers 140°F. Using a slotted spoon or spatula, transfer the salmon to a platter, cover, and refrigerate for 1 to 2 hours, or until chilled.

3. To serve, toss the spinach with the oil and vinegar. Arrange the spinach on each of 6 plates. Set a fillet on top of the greens and spoon the cucumber-dill sauce over the top. Garnish each plate with 2 or 3 rounds of cucumber or zucchini and a cherry tomato. Serve with whole grain bread or crackers, if desired.

SWEET POTATO OVEN FRIES

SERVES 4

Enough sprigs of fresh rosemary to cover a baking sheet
1 teaspoon chili powder
1 teaspoon ground cumin
1 teaspoon paprika
1 teaspoon kosher salt
1 teaspoon freshly ground black pepper

2 medium sweet potatoes (about 1 pound), scrubbed and blotted dry

1. Preheat the oven to 400°F. Spray a large baking sheet with olive oil spray, or line it with parchment paper, or use an ungreased, unlined nonstick baking sheet.

2. Spread the rosemary sprigs on the baking sheet in a single layer, making sure that the entire surface is covered.

3. In a small bowl, mix together the chili powder, cumin, paprika, salt, and pepper.

4. Square off the potatoes by slicing off the sides lengthwise about ½ inch in from the edge. Cut crosswise into ½-inch strips the size of steak fries, leaving the skin on. Slice the remaining rectangles into ½-inch strips, also the size of steak fries.

5. Lay the potato strips on the rosemary sprigs in a single layer and sprinkle generously with the seasoning mixture. Spray generously with olive oil spray.

6. Bake for 20 minutes, remove from the oven, and spray again with olive oil spray.

7. Return to the oven for about 25 minutes, or until the fries are golden and puffed. Brush off the rosemary sprigs and serve warm.

■ *Dinner*

STIR-FRIED TOFU WITH SOBA NOODLES

SERVES 6

Tofu
    2 teaspoons extra virgin olive oil
    1 to 3 garlic cloves, minced

1 tablespoon minced or grated fresh ginger

2 large red bell peppers, cut into pieces about 2 inches long and ¼ inch wide (about 1 cup)

¼ teaspoon freshly ground black pepper

Pinch of red pepper flakes

1 tablespoon finely chopped fresh basil or 1 teaspoon dried basil

2 tablespoons low-sodium soy sauce

1 pound extra-firm tofu, cut into ½-inch cubes

Noodles

6 ounces soba (buckwheat) noodles

1 tablespoon olive oil

2 to 3 garlic cloves, minced

3 tablespoons sesame seeds

1 to 2 tablespoons minced fresh cilantro

¼ teaspoon freshly ground black pepper

1. To prepare the tofu, in a nonstick skillet, heat the olive oil over medium-high heat and cook the garlic, ginger, and red bell peppers, stirring, for about 5 minutes or until the peppers begin to soften. Add the black pepper, pepper flakes, basil, soy sauce, and tofu and stir to coat the tofu. Cook for 10 to 12 minutes, or until heated through.

2. To prepare the noodles, cook the soba noodles in boiling water for 5 to 7 minutes, or until tender. Drain and rinse well under cool running water. Set aside.

3. In a large nonstick skillet, heat the oil over medium-high heat. Add the garlic and cook, stirring, for about 30 seconds, or until tender. Add the sesame seeds and cook for about 30 seconds until fragrant, stirring constantly to prevent burning. Stir in the cilantro and pepper.

4. Add the noodles to the skillet and toss well. Spoon

equal portions of noodles into 6 bowls, divide the tofu mixture between them, and serve.

GOLDEN DOOR BLUEBERRY BREAD

MAKES 1 LOAF

1 cup unbleached, all-purpose flour
1 teaspoon baking powder
½ teaspoon baking soda
2 teaspoons ground cinnamon
½ teaspoon ground allspice
1¼ cups whole wheat flour
¼ cup stone-ground yellow cornmeal
½ teaspoon kosher salt
2 ripe medium bananas, mashed (about 1 cup)
½ cup light brown sugar
1 large omega-3 egg
1 large omega-3 egg white
2 tablespoons canola oil
1½ cups low-fat buttermilk
2 tablespoons grated orange zest
1 cup chopped walnuts
1 cup blueberries, soaked in warm water for about
    15 minutes, and drained
Nonfat vanilla frozen yogurt, optional

1. Preheat the oven to 350°F. Spray a 9 x 5-inch or 8½ x 4-inch loaf pan with vegetable oil spray.

2. In a large mixing bowl, sift together the all-purpose flour, baking powder, baking soda, cinnamon, and allspice. Stir in the whole wheat flour, cornmeal, and salt.

3. In a blender or food processor fitted with a metal blade, combine the bananas, brown sugar, egg, egg white, canola oil, and buttermilk, and process until smooth. Add the orange zest and pulse until just combined.

4. Make a well in the flour mixture. Pour the banana mixture into it and mix until almost incorporated. Add the walnuts and drained blueberries and gently mix into the batter. Do not overmix.

5. Pour the batter into the prepared pan and bake for about 55 minutes, or until a toothpick inserted into the center comes out clean, the top is golden brown, and the bread begins to pull away from the sides of the pan. Remove from the oven, invert on a wire rack, and cool briefly before slicing.

6. Slice into ½- to ¾-inch-thick slices. Serve each slice topped with ½-cup scoop of nonfat vanilla frozen yogurt, if desired.

DAY 4

MENU

Breakfast: Sweet Potato Pie
Morning Snack: 8 ounces 100% pomegranate juice (mixed with sparkling water if desired)
Lunch: Tuna Salad with Mesclun, Basil, and Sprouts
Afternoon Snack: 1 cup papaya chunks, 8 ounces Knudsen Very Veggie Cocktail, 1 ounce unsalted dry-roasted or raw almonds
Dinner: Asian Stir-Fry with Turkey Cutlets, Basmati Rice, Wilted Greens and Peanut-Ginger Sauce; SuperFoods Rx Salad (page 273); No-Bake Pumpkin-Yogurt Cheese Pie

SWEET POTATO PIE

SERVES 8

Pie for breakfast? You bet. Make this the night before. Refrigerate once it has cooled.

Crust
- 1½ cups crushed whole wheat, low-fat gingersnap cookies, see Notes
- 3 tablespoons ground flaxseed or wheat germ
- 1 tablespoon dark brown sugar
- 1 large egg

Filling
- 2 garnet yams, see Notes
- 1 navel orange
- 1 ripe banana
- ½ cup fat-free ricotta cheese or light silken tofu
- 2 large omega-3 eggs
- 2 teaspoons pure vanilla extract
- 1 teaspoon ground cinnamon
- ¼ teaspoon ground nutmeg or mace
- ¼ teaspoon ground cloves
- ⅓ to ½ cup pure maple syrup, or to taste

Garnish
- Vanilla yogurt cheese, optional (see page 240)
- 8 raspberries or strawberries
- Mint sprigs

1. Preheat the oven to 350°F.

2. To make the crust, put the gingersnap crumbs, flaxseed, and brown sugar in a food processor fitted with a metal blade or a blender and process to a powder. Add the egg and blend until moist.

3. Press the crumb mixture into a lightly oiled 9-inch pie plate, gently pushing it across the bottom and up the sides. Bake for about 20 minutes, or until the crust is golden brown. Cool the crust completely on a wire rack. Do not turn off the oven.

4. Meanwhile, pierce the yams in a few places with the tines of a fork or a sharp knife. Bake for about 1 hour and 15 minutes, or until tender when pierced with a knife. Set aside to cool. Do not turn off the oven.

5. When cool enough to handle, peel the yams and mash the flesh. Measure 2 cups of mashed yams.

6. Slice the top and bottom off the orange but do not peel it. Cut the orange into quarters and put it in a food processor fitted with a metal blade or a large blender. Add the mashed yams, banana, ricotta, eggs, vanilla, cinnamon, nutmeg, and cloves. Blend until smooth. Add ⅓ cup of the syrup, blend, and taste. Add more syrup if necessary.

7. Scrape the filling into the cooled crust and bake on the center rack of the oven for 40 to 50 minutes, or until set in the center. If the crust darkens during baking, shield the edges with strips of foil.

8. Cool the pie on a wire rack. Cut into wedges and serve with dollops of yogurt cheese, berries, and mint sprigs.

NOTES: We recommend garnet or red yams, which are sweeter and creamier than jewel yams. All tubers labeled "yams" in our American markets are actually sweet potatoes.

For the crust, you may substitute graham cracker crumbs mixed with 1 teaspoon of ground ginger for the gingersnap crumbs. Eight graham cracker squares yield 1 cup of crumbs.

■ *Lunch*

TUNA SALAD WITH MESCLUN, BASIL, AND SPROUTS

SERVES 4

Two 13-ounce cans water-packed albacore tuna, drained
2 ribs celery, cut into ¼-inch dice
½ cup diced red onion
1 cup nonfat plain yogurt
2 tablespoons Dijon mustard
2 teaspoons rice vinegar
2 teaspoons dried dill
8 cups mixed mesclun greens, spinach, or romaine, washed and torn into bite-sized pieces
½ cup torn fresh basil leaves
¼ cup alfalfa sprouts
¼ cup radish sprouts
¼ cup Raspberry Vinaigrette (page 254)
⅛ avocado, sliced
1 large hard-cooked omega-3 egg, sliced
½ yellow bell pepper, sliced
4 ripe medium tomatoes, cut into wedges
Freshly ground black pepper
¼ cup edible flowers, such as nasturtiums, geraniums, or petunias (optional)

1. Flake the tuna into a small bowl. Add the celery, onion, yogurt, mustard, vinegar, and dill and mix well.

2. In a large bowl, toss the greens, basil, and sprouts with the vinaigrette.

3. To compose the salad, arrange the tossed greens on four large plates. Add equal portions of avocado, egg, bell pepper, and tomatoes to each serving. Using an ice cream scoop, portion the tuna salad in the center of the greens. Garnish with fresh pepper and edible flowers, if desired.

■ Dinner

ASIAN STIR-FRY WITH TURKEY CUTLETS, BASMATI RICE, WILTED GREENS, AND PEANUT-GINGER SAUCE

SERVES 4

Turkey and marinade
  1 teaspoon minced fresh ginger
  ½ teaspoon toasted sesame oil
  2 teaspoons low-sodium tamari or soy sauce
  1 tablespoon fresh lime juice
  1 teaspoon cracked Szechuan peppercorns or chile paste
  1 pound skinless, boneless turkey cutlets or breast

Basmati rice
  ¾ cup brown basmati rice
  1¾ cups vegetable stock or water

Peanut-ginger sauce
  2 tablespoons finely chopped fresh ginger
  2 to 3 tablespoons low-sodium tamari

2 tablespoons rice vinegar

1 tablespoon red wine vinegar

2 tablespoons honey

1 ½ teaspoons dried basil

2 tablespoons crunchy peanut butter

2 tablespoons water

Vegetable stir-fry

¼ teaspoon canola oil

1 medium carrot, thinly sliced on the diagonal

1 celery rib, thinly sliced on the diagonal

8 shiitake mushrooms, thinly sliced

¼ cup vegetable stock or water

Wilted greens

¼ teaspoon canola oil

1 teaspoon minced garlic

1 teaspoon minced fresh ginger

¼ teaspoon fennel seeds, crushed

1 ½ small fennel bulbs, thinly sliced (about 1 ½ cups)

2 to 3 tablespoons vegetable or chicken stock or water

1 cup torn Swiss chard

1 cup fresh spinach leaves

2 cups torn bok choy leaves

2 teaspoons low-sodium soy sauce

½ teaspoon toasted sesame oil

1. To marinate the turkey, in a shallow glass or ceramic dish, whisk together the ginger, sesame oil, tamari, lime juice, and cracked peppercorns. Add the turkey, turn to coat, cover, and refrigerate for at least 1 hour and up to 8 hours.

2. To make the basmati rice, combine the rice and stock in a medium saucepan. Bring to a boil over high heat,

reduce the heat, cover and simmer for about 30 minutes or until the liquid is absorbed. Remove from the heat and let stand for 5 minutes. Remove the cover and fluff the rice with a fork. Cover and keep warm.

3. To make the peanut sauce, put the ginger, tamari, rice vinegar, red wine vinegar, honey, basil, peanut butter, and water into a blender and process until smooth. Set aside.

4. Preheat the oven to 350°F. Spray an ovenproof, non-stick skillet pan with vegetable oil spray.

5. Remove the turkey from the marinade and discard the marinade. Heat the oiled pan over medium-high heat, add the turkey cutlets and sear for about 1 minute, or until golden. Using tongs or a spatula, turn and sear the other side for about 1 minute. Partially cover the pan and bake for 10 to 15 minutes, or until the turkey is no longer pink in the center but still juicy. An instant-read thermometer inserted into the thickest part of the turkey should register 170°F. Transfer to a warmed platter.

6. To stir-fry the vegetables, heat the oil in a wok or large nonstick sauté pan over medium-high heat. Add the carrot and celery and stir-fry for 2 to 3 minutes, or until they begin to soften. Add the mushrooms and cook for 1 minute longer. Add the stock or water to deglaze the pan, loosening any browned bits from the bottom with a wooden spoon. Transfer the vegetables to a bowl and keep warm.

7. To sauté the greens, in the same wok or sauté pan, heat the oil over medium-high heat. Add the garlic, ginger, fennel seeds, and fennel and sauté for 1 minute. If the vegetables begin to stick, deglaze the pan with a drizzle of vegetable stock or defatted chicken stock. Add the Swiss chard, spinach, and bok choy and continue cooking just until the greens wilt. Remove from the heat and toss in the soy sauce and sesame oil.

8. Layer equal portions of basmati rice, wilted greens,

and vegetables on each of 4 plates. Top each with turkey and drizzle with peanut sauce.

SUPERFOODS RX SALAD

SERVES 1

I eat this salad almost every day—sometimes twice a day. Feel free to vary the ingredients using the *SuperFoods Rx* sidekicks. Add a handful of your favorite chopped fresh herb, 1 tablespoon grated Parmigiano-Reggiano, or 2 tablespoons roasted nuts or seeds. The quantity is for one person, but you can easily multiply the ingredients to serve more.

   1 cup spinach torn into bite-sized pieces
   1 cup chopped romaine
   ¼ cup shredded red cabbage
   ½ cup sliced red bell pepper
   ½ tomato, chopped
   ¼ cup chickpeas; if canned, rinse well
   ½ cup grated carrot
   ¼ avocado, cubed
   2 tablespoons extra virgin olive oil
   1 teaspoon balsamic vinegar

Combine the spinach, romaine, cabbage, red pepper, tomato, chickpeas, carrot, and avocado in a bowl. In a separate bowl, whisk together the olive oil and vinegar. Toss the dressing with the salad just before serving.

NO-BAKE PUMPKIN-YOGURT CHEESE PIE

Crumb crust

1 cup whole wheat graham cracker crumbs (from 8 graham cracker squares)

½ cup whole wheat pastry flour

2 tablespoons canola or sunflower oil

2 tablespoons dark brown sugar

2 tablespoons ground flaxseed

2 tablespoons wheat germ or ground pumpkin seeds

2 teaspoons finely grated orange zest

½ teaspoon ground ginger

1 large omega-3 egg, lightly beaten

Filling

24 ounces canned plain pumpkin puree or 20 cups cooked sugar pumpkin or butternut squash (about 4 pounds uncooked squash)

1 cup nonfat plain yogurt cheese or 1 cup fat-free ricotta, see Note

¾ to 1 cup pure maple syrup

Two ¼-ounce packets gelatin

1 teaspoon ground cinnamon

¼ teaspoon ground ginger

¼ teaspoon ground cloves

¼ teaspoon ground allspice

Mint sprigs, for garnish

Fresh berries, for garnish, optional

1. Preheat the oven to 350°F.

2. To make the crust, mix together the graham cracker crumbs, flour, oil, brown sugar, flaxseed, wheat germ, zest,

and ginger. Toss the mix. Add the egg and stir until the mixture is moistened.

3. Press the crumb mixture into a lightly oiled 9-inch pie plate, gently pushing it over the bottom and up the sides. Bake for about 20 minutes, or until the crust is golden brown. Cool the crust completely on a wire rack.

4. To make the filling, put the pumpkin puree and yogurt cheese in a food processor fitted with a metal blade or a blender and puree until smooth.

5. In a small saucepan, heat the syrup over medium heat until simmering. Remove from heat, add the gelatin, cinnamon, ginger, cloves, and allspice and stir until dissolved. Pour most of the syrup into the pumpkin mixture and puree until incorporated. Taste and add the rest of the syrup, if needed.

6. Carefully pour into the crust. Cover with plastic wrap and chill for at least 2 hours or until set.

7. To serve, cut into 8 wedges. Garnish each slice with a sprig of mint and a berry or two, if desired.

NOTE: To make yogurt cheese, put 1 quart of nonfat plain yogurt in a stainless steel or plastic mesh strainer lined with a large unbleached coffee filter and set over a bowl. Put the bowl in the refrigerator and let the yogurt drain for 6 to 8 hours. You will have 2 cups of yogurt cheese. Reserve the drained whey to use when baking muffins or making pancakes.

DAY 5

MENU

Breakfast: Breakfast Broccoli Frittata, 4 ounces pink grapefruit juice

Morning Snack: ¼ cup canned pumpkin mixed with ½ cup unsweetened applesauce

Lunch: Turkey Burger with the Works, Vegetable Stir-Fry, 8 ounces low-sodium tomato juice

Afternoon Snack: ¼ medium cantaloupe, cubed, mixed with 1 cup nonfat yogurt, 1 tablespoon flaxmeal, and 2 tablespoons wheat germ

Dinner: White Bean Soup with Greens and Rosemary, Tomato and White Wine–Braised Halibut, Dried Apricot and Cranberry Compote with Apples

■ *Breakfast*

BREAKFAST BROCCOLI FRITTATA

SERVES 4

2 medium potatoes, sliced into ⅛-inch-thick rounds
1½ cups broccoli florets
½ medium onion, diced
½ red bell pepper, seeded, ribs removed, and chopped
½ yellow bell pepper, seeded, ribs removed, and chopped
4 large omega-3 eggs
1 cup 1% cottage cheese or fat-free ricotta
½ tablespoon fresh dill or ¾ teaspoons dried dill
3 tablespoons Asiago or Parmesan cheese
4 slices whole grain bread, optional
Tomato slices, for garnish, optional
Melon slices, for garnish, optional

1. Preheat the oven to 350°F. Spray a pie plate with olive oil spray.

2. Lay the potato slices in a single layer in the bottom of the pie plate. Cut the remaining slices in half and arrange them around the sides of the pie plate. Bake for 12 to 15 minutes, or until lightly browned.

3. In a large sauté pan sprayed with olive oil spray, sauté the broccoli, onion, and bell peppers for 3 to 4 minutes, or until tender. Add a tablespoon of water or vegetable stock if needed to prevent sticking.

4. In a mixing bowl, beat together the eggs and cottage cheese. Stir in the sautéed vegetables, dill, and half of the grated cheese. Add the egg mixture to the pie plate and bake for 20 to 25 minutes, or until the egg is set. Sprinkle with the remaining grated cheese and return the frittata to the oven for a minute or two to allow the cheese to melt. Cut into wedges to serve. Serve each wedge with a slice of bread, tomato, or melon, if desired.

■ *Lunch*

TURKEY BURGERS WITH THE WORKS
··

SERVES 4

Burgers
   ¾ pound lean, ground, skinless turkey breast and thigh
      meat, see Note
   1½ teaspoons canola or safflower oil
   1½ tablespoons minced shallots
   1 tablespoon ground flaxseed
   1 tablespoon whole wheat flour
   1 tablespoon minced fresh flat-leaf parsley
   2 large egg whites
   Freshly ground black pepper

The works
    4 whole grain buns
    Dijon mustard
    8 large romaine or red lettuce leaves
    2 ripe tomatoes, cut into ¼-inch-thick slices
    ½ medium red onion, cut into ⅛-inch-thick slices
    ½ medium avocado, sliced in wedges
    Ketchup or chunky salsa

1. To make the burgers, in a bowl and using your hands or a large spoon, mix together the turkey, oil, shallots, flaxseed, flour, parsley, egg whites, and pepper. Gently form the mixture into 4 patties. Do not pack the meat tightly. Transfer to a tray, cover, and refrigerate until ready to grill.

2. Prepare a charcoal or gas grill so that the coals are medium-hot to hot. Spray the grilling rack with vegetable oil spray and return the rack to the grill.

3. Grill the burgers for 5 minutes on one side, turn and grill for 8 to 10 minutes longer, or until the burgers are cooked through and an instant-read thermometer inserted in the thickest part of the burger registers 170°F. and the juices run clear.

4. Meanwhile, toast the buns, cut side down, on the edge of the grill just until lightly browned. Spread the bottom half of each bun with mustard.

5. Divide the lettuce leaves, tomato and onion slices, and avocado wedges among the buns. Top each with a burger and spread ketchup or salsa on the burgers. Serve immediately.

NOTE: The best way to ensure that you get quality ground turkey is to grind your own or to ask the butcher to grind white- and dark-meat turkey for you. If you buy turkey in the markets, buy ground turkey

breast so you know exactly what you are getting. Turkey is leaner than beef, so it needs the oil and egg whites to keep it moist and the flour for binding.

VEGETABLE STIR-FRY

SERVES 4

1 teaspoon canola oil
2 teaspoons minced garlic
½ teaspoon minced fresh ginger
12 asparagus stalks, trimmed and sliced diagonally into 2-inch pieces
1 medium carrot, sliced diagonally into ⅛-inch pieces
One 5-ounce can water chestnuts, drained and sliced into ⅛-inch rounds
6 scallions, including tops, sliced diagonally into ½-inch pieces
4 teaspoons low-sodium tamari
4 tablespoons vegetable stock or water
2 teaspoons cornstarch dissolved in 2 tablespoons water
2 teaspoons sesame seeds

1. In a nonstick saucepan or wok, heat the oil over medium-high heat. Add the garlic and ginger, and cook for about 1 minute, stirring, or until softened.

2. In quick succession, add the asparagus, carrot, and water chestnuts, and stir-fry for 2 to 3 minutes, or until the vegetables begin to soften. Add the scallions and cook for 1 minute longer.

3. Add the tamari and toss gently with a wooden spoon to distribute the liquid among the vegetables. Add the stock and toss again. Add the dissolved arrowroot and toss once

more. Cook for 2 to 3 minutes, or until the liquid comes to a simmer and starts to thicken. Sprinkle with sesame seeds and serve.

■ *Dinner*

WHITE BEAN SOUP WITH GREENS AND ROSEMARY

SERVES 6

1 cup dried white beans
4 cups vegetable stock or water
1 bay leaf
1 tablespoon olive oil
2 medium carrots, cubed
1 medium onion, cubed
2 garlic cloves, minced
2½ tablespoons low-sodium soy sauce
1 tablespoon chopped fresh rosemary leaves
1 teaspoon chopped fresh thyme
¼ teaspoon freshly ground black pepper
Pinch of cayenne
1 bunch spinach, kale, or chard (10 to 12 ounces), rinsed and stemmed
3 tablespoons grated Parmesan cheese, optional

1. To prepare the beans, put them in a large bowl and add enough cold water to cover by one or two inches. Set aside to soak for at least 6 hours and up to 12 hours. Change the water 2 to 3 times during soaking. Drain the beans.

2. In a stockpot, combine the drained beans, stock, and bay leaf and bring to a boil over high heat. Reduce the heat, cover, and simmer gently for about 1 hour and 30 minutes, or until the beans are tender but not mushy.

3. Meanwhile, in a large sauté pan, heat the oil over medium heat and sauté the carrots and onion for about 1 minute. Add the garlic and cook for about 5 minutes, or until softened. Transfer to the pot with the beans.

4. Cook the soup for about 15 minutes over medium-low heat. Add the soy sauce, rosemary, thyme, pepper, and cayenne and cook for about 15 minutes longer, or until the flavors meld. Add the spinach and cook for about 5 minutes, or just until wilted. (If using kale, allow 10 minutes for it to wilt.)

5. Ladle the soup into bowls and sprinkle with cheese, if desired, before serving.

TOMATO AND WHITE WINE–BRAISED HALIBUT

SERVES 4

2 tablespoons olive oil
½ medium onion, finely diced
¾ cup finely diced celery
¾ cup finely diced carrot
2 garlic cloves, minced
1 cup white wine
1 tablespoon tomato paste
4 cups chicken or vegetable stock
1 cup crushed tomatoes
¼ cup finely chopped fresh flat-leaf parsley
2 tablespoons finely chopped fresh thyme
1 teaspoon grated lemon zest

1 teaspoon grated orange zest
Four 4- or 5-ounce halibut, mahi mahi, or sea bass fillets
½ teaspoon kosher salt
1 teaspoon freshly ground black pepper

1. Preheat the oven to 375°F.
2. In a large saucepan, heat the olive oil over medium-high heat and cook the onion, celery, carrot, and garlic for 3 to 4 minutes, or until golden brown. Stir in the wine and tomato paste, bring to a simmer, and cook for 10 to 12 minutes, or until the sauce is reduced by half.
3. Add the stock and crushed tomatoes and simmer the sauce for 30 to 35 minutes, or until reduced to about 3 or 4 cups. Transfer to a glass or ceramic dish, cover to keep warm, and set aside.
4. In a small bowl, mix together the parsley, thyme, lemon zest, and orange zest. Refrigerate until ready to use.
5. Season the halibut fillets with salt and pepper. Arrange the fillets on top of the tomato sauce, cover, and bake for 20 to 25 minutes, or until the center of the fish is white but not transparent. Serve the fish with some sauce and a sprinkling of the parsley mixture.

DRIED APRICOT AND CRANBERRY COMPOTE WITH APPLES

SERVES 8

1½ cups dried apricots
¾ cup dried cranberries, dried cherries, currants, or raisins
2 medium apples, peeled, cored, and cut into large pieces
¼ cup unsweetened apple juice
¾ cup light brown sugar

½ cinnamon stick
2 tablespoons julienned orange or lemon zest
Nonfat plain or vanilla yogurt or nonfat frozen vanilla yogurt,
  optional

1. In a small bowl, combine the dried apricots and cranberries and add enough warm water to cover. Soak for 30 minutes. Drain and discard the soaking liquid.

2. Transfer the apricots and cranberries to a medium-size saucepan. Add the apples, apple juice, brown sugar, and cinnamon stick and bring to a simmer over medium heat. Cover and simmer for 10 to 15 minutes, or until the fruit begins to break down. Uncover the pan and simmer for a few minutes longer, breaking up the fruit with a wooden spoon, until slightly thickened and chunky. Remove the cinnamon stick and set the fruit aside to cool.

3. Meanwhile, in a small saucepan, bring 1 or 2 inches of water to a boil over medium-high heat. Add the orange or lemon zest and blanch for about 45 seconds. Drain and set aside to cool.

4. Spoon the cooled, thickened fruit into small bowls. Garnish with the cooled zest. Serve with a dollop of yogurt, if desired.

DAY 6

MENU

Breakfast: Rancho La Puerta Breakfast Crisp, 1 cup papaya chunks, 8 ounces soymilk
Morning Snack: 2 cups watermelon cubes
Lunch: Braised Tofu-Stuffed Whole Wheat Pita Sandwich, Pumpkin Soup

Afternoon Snack: ½ medium orange bell pepper, cut into strips; 8 ounces low-sodium tomato juice

Dinner: Baked Turkey with Curried Currant Sauce and Wilted Spinach, Ricotta Torte with Blueberries

■ *Breakfast*

RANCHO LA PUERTA BREAKFAST CRISP

SERVES 8

Sip a 4-ounce glass of unsweetened grape juice with this.

Crisp

- ⅓ cup chopped almonds
- ⅓ cup chopped pecans
- ⅓ cup chopped walnuts
- 1 cup quick-cooking or regular rolled oats
- 3 to 4 tablespoons pure maple syrup
- 2 tablespoons wheat germ
- 2 tablespoons whole wheat pastry flour or unbleached all-purpose flour
- ½ teaspoon pure vanilla extract
- ½ teaspoon ground cinnamon
- ¼ teaspoon ground mace or nutmeg

Filling

- 3 to 4 pears, apples, peaches, or nectarines, peeled, pitted or cored, and sliced
- ½ cup dried berries, cherries, currants, or raisins
- ½ teaspoon ground cinnamon
- ¼ teaspoon ground mace or nutmeg
- 2 teaspoons pure maple syrup

Topping

16 ounces nonfat vanilla yogurt

1. To prepare the crisp, preheat the oven to 325°F.

2. Spread the almonds, pecans, and walnuts on a baking sheet and roast for 5 to 7 minutes, or until golden brown and fragrant. Remove from the oven and slide onto a plate to halt the cooking.

3. In a mixing bowl, combine the nuts, oats, syrup, wheat germ, flour, vanilla, cinnamon, and mace and toss to mix evenly.

4. For the filling, cut the fruit slices into small pieces. Transfer to a mixing bowl and add the dried berries, cinnamon, and mace. Add the syrup and toss until evenly mixed. Spread the fruit in a square 8-inch baking pan or pie plate. Top the fruit with the crisp topping.

5. Bake for 15 to 20 minutes, or until the fruit is tender and the top lightly browned. Serve warm, topped with dollops of yogurt.

▪ Lunch

BRAISED TOFU-STUFFED WHOLE WHEAT PITA SANDWICH

SERVES 4

Roasted Red Pepper Hummus (page 286)
2 medium tomatoes, sliced
Braised Tofu (page 287)
½ cup bean sprouts
½ cup mixed salad greens
½ cup shredded carrots

½ avocado, sliced into wedges
Four 8-inch whole wheat pitas, halved

Spread 1½ tablespoons of hummus on one half of a pita. Top with tomato slices, 2 slices of tofu, sprouts, salad greens, carrots, and avocado slices. Press closed and serve. Repeat to make 3 more sandwiches.

ROASTED RED PEPPER HUMMUS

SERVES 8; MAKES ABOUT 5 CUPS

1 cup dried garbanzo beans
6 cups vegetable stock or water
2 bay leaves
1 teaspoon cumin seed or ½ teaspoon ground cumin
1 medium red bell pepper
½ cup nonfat plain yogurt or ¼ cup soft silken tofu
¼ cup fresh lemon juice
¼ cup orange juice
2 tablespoons tahini
1 tablespoon olive oil
3 to 5 garlic cloves, minced
½ teaspoon sea salt, or to taste

Garnish

2 to 3 scallions, white and green parts, thinly sliced
¼ cup minced fresh flat-leaf parsley
1 tablespoon finely chopped orange zest, optional
Sliced red or yellow bell peppers, carrots, zucchini, and
    broccoli, for dipping
Whole wheat pita bread, cut into triangles

1. To prepare the beans, put them in a large bowl and add enough cold water to cover by one or two inches. Set aside to soak for at least 6 hours and up to 12 hours. Change the water 2 to 3 times during soaking. Drain the beans.

2. In a stockpot, combine the drained beans, stock, bay leaves, and cumin and bring to a boil over high heat. Reduce the heat, cover, and simmer gently for about 1 hour and 30 minutes to 2 hours, or until the beans are tender but not mushy. Drain.

3. Preheat the oven to 350°F.

4. Cut the red bell pepper in half, remove the stem, scrape out the seeds, and lay, cut side down, on a baking sheet. Bake until the pepper is tender and the skin begins to wrinkle and color. Using the dull side of a paring knife, peel off the skin and cut the pepper halves into chunks.

5. Put the beans, bell peppers, yogurt, lemon and orange juices, tahini, oil, and garlic in the bowl of a food processor fitted with a metal blade. Process until smooth. Add the salt, pulse to mix, taste and adjust the seasoning.

6. Spoon into a serving bowl and garnish with green onions, parsley, and zest, if desired. Serve with sliced raw vegetables and pita triangles.

BRAISED TOFU

MAKES 1 POUND; ABOUT 2 CUPS

1 pound fresh firm tofu, frozen and then defrosted, see Note

1 teaspoon extra virgin olive oil, see Note

8 garlic cloves, coarsely chopped

1 tablespoon low-sodium soy sauce
1 tablespoon fresh lemon juice, balsamic vinegar, or
    Worcestershire sauce

1. Cut the defrosted tofu into ½-inch cubes.
2. In a skillet, heat the olive oil over medium heat and cook the garlic for about 1 minute. Add the soy sauce, lemon juice, and tofu, reduce the heat to low, and cook, turning occasionally, for 20 to 30 minutes, or until the tofu is browned.

NOTE: Freezing changes the texture of tofu, giving it a meaty "mouthfeel." Drain fresh firm tofu and cut into 8 slices. Double wrap the slices in plastic wrap. Freeze for at least 8 hours and up to several weeks. Defrost the tofu slowly in the refrigerator, or to hasten the process, submerge the still-wrapped tofu in a pot filled with boiling water. Remove the pot from the heat and let the tofu sit in the hot water for 10 to 15 minutes. Lift the tofu from the water, unwrap, and proceed with the recipe.

To give the tofu an Asian tang, substitute sesame oil for the olive oil and add 2 tablespoons of grated fresh ginger with the garlic.

PUMPKIN SOUP

SERVES 8

2 teaspoons olive oil
2 cups leeks, white parts only, washed well and cut into
    1-inch slices
½ medium yellow onion, coarsely chopped
2 medium carrots, cut into 1-inch slices

1 medium green apple, peeled, cored, and cut into 1-inch
   chunks
Two 16-ounce cans unseasoned pumpkin puree
1 sprig fresh thyme or ½ teaspoon dried thyme
1 bay leaf
2 teaspoons salt
1 teaspoon freshly ground black pepper, or to taste
1 teaspoon ground allspice
1 teaspoon ground cinnamon
7 cups vegetable stock or water
¼ cup frozen orange juice concentrate, thawed
3 tablespoons nonfat yogurt, for garnish
1 medium green apple, peeled, cored, and thinly sliced, for
   garnish

1. In a large soup pot, heat the oil over medium heat.
Add the leeks, onion, carrots, apple, pumpkin, thyme, bay
leaf, salt, pepper, allspice, and cinnamon. Cover and cook
for about 10 minutes, stirring once or twice and adjusting
the heat if necessary, until vegetables begin to soften.

2. Stir in the stock, bring to a boil, reduce the heat to
medium-low, and simmer, uncovered, for about 45 minutes,
or until the vegetables are tender. Let cool for about 15
minutes. Discard the thyme sprig and bay leaf.

3. Working in batches, transfer to a blender or food
processor fitted with a metal blade and process until
smooth.

4. Pour the soup back into the pot, and cook over
medium heat until hot and the flavors blend. Season with
salt and pepper, if necessary, and stir in the orange juice
concentrate. Add more vegetable stock if necessary to
obtain the desired consistency.

5. Ladle the hot soup into bowls and garnish with dollops of yogurt and sliced apples.

BAKED TURKEY WITH CURRIED CURRANT SAUCE AND WILTED SPINACH

SERVES 4

Turkey and sauce

3 pounds boneless, skinless turkey breasts
1 pound carrots
1 white onion, cut in half
2 sprigs fresh thyme
1 bay leaf
2 teaspoons freshly ground black pepper
1 cup vegetable stock or water, or more as needed
1 teaspoon cornstarch
1 tablespoon curry powder
¾ cup currants, plumped in warm water and drained

Wilted spinach

1 teaspoon canola oil
2 teaspoons minced garlic
4 cups fresh spinach leaves
Minced fresh flat-leaf parsley, for garnish
⅔ cup long-grain brown rice, cooked
2 tablespoons flaxmeal

1. To prepare the turkey, preheat the oven to 350°F.

2. Put the turkey in a deep roasting pan that is just large enough to contain it. Scatter the carrots, onion, thyme, and bay leaf around the turkey. Season with pepper. Cover with foil to seal it. Bake for about 1 hour and 45 minutes until the juices run clear and an instant-read thermometer registers 170°F. when inserted in the thickest part of the breast.

3. Remove from the oven and let rest for 15 minutes. Remove the breast from the pan and slice thinly. Reserve the pan's contents.

4. Discard the bay leaf and thyme sprigs from the pan. Using a slotted spoon, transfer the carrots and onion to a blender.

5. Skim as much fat as possible from the pan juices or blot them with paper towels to remove the fat. Pour the defatted juices into a measuring cup, add enough vegetable stock to measure 2 cups, and pour into the blender. Add the cornstarch and curry powder, and blend at high speed until smooth.

6. Transfer to a small saucepan, add the drained currants, and cook over low heat until heated through and the currants begin to pop. Season with black pepper to taste, if necessary. Keep warm until ready to serve.

7. To prepare the spinach, in a large sauté pan, heat the oil over medium-high heat and cook the garlic, stirring, for 1 minute, or until softened but not browned. Add the spinach and continue cooking just until the greens wilt. Remove from heat and keep warm.

8. Toss the brown rice with the flaxmeal and divide among 4 warmed plates. Spoon 1 cup wilted spinach and fan several slices of turkey on each plate. Ladle 3 tablespoons warm sauce over the turkey. Sprinkle with minced parsley and serve.

RICOTTA TORTE WITH BLUEBERRIES

SERVES 8

Crust

3 tablespoons flaxseed
3 tablespoons toasted almonds
2 large whole wheat graham crackers
2 tablespoons light brown sugar
¼ teaspoon almond extract

Filling

4 large omega-3 eggs
1 pound nonfat ricotta cheese (about 2 cups)
3 tablespoons whole wheat pastry flour or unbleached flour
⅓ cup granulated sugar
1 teaspoon pure vanilla extract
1 tablespoon finely grated lemon zest
3 cups fresh or frozen blueberries (thawed, if frozen)
Confectioners' sugar

1. Preheat the oven to 325°F. Butter or oil a 1-quart soufflé pan.

2. In a food processor fitted with a metal blade, process the flaxseed and almonds until finely ground. Add the graham crackers, brown sugar, and almond extract, and process again until ground to fine crumbs and well mixed. Press the mixture into the prepared soufflé pan, gently pushing it over the bottom and up the sides

3. Wipe out the food processor and add the eggs, ricotta, flour, granulated sugar, and vanilla, and process until smooth. Or whisk by hand in a medium bowl. Stir in the lemon zest and carefully pour the batter into the soufflé pan.

4. Bake for about 50 minutes, or until the torte is set and a toothpick inserted into the center comes out clean. Cool on a wire rack for about 20 minutes.

5. To serve, run a knife blade around the edge of the pan. Slice into wedges, serve with berries, and garnish with confectioners' sugar.

DAY 7

MENU

Breakfast: Sweet Potato Scones, Grape Cooler
Morning Snack: 1 ounce walnuts, 8 ounces Knudsen Very Veggie Cocktail
Lunch: Crab-Stuffed Papaya; Spinach, Mushroom, and Egg Chopped Salad
Afternoon Snack: 1 medium carrot, 1 ounce unsalted soy nuts
Dinner: Orange-Ginger Chicken with Apricot-Almond Couscous, Sesame Stir-Braised Kale, ½ cup vanilla frozen yogurt

■ *Breakfast*

SWEET POTATO SCONES

MAKES ABOUT 18 SCONES

1 cup plus 2 tablespoons whole wheat pastry flour
¼ cup oat or wheat bran
2 tablespoons ground flaxseed
2 tablespoons wheat germ
2 teaspoons baking powder

½ teaspoon baking soda

¼ teaspoon ground cinnamon

⅛ teaspoon ground mace or nutmeg

2 tablespoons dark brown sugar

⅓ cup dried blueberries, currants, or chopped raisins

1 cup shredded sweet potato or winter squash, such as butternut, acorn, or Hubbard

⅓ cup nonfat yogurt or buttermilk

1 large omega-3 egg

1½ tablespoons canola or safflower oil

2 teaspoons finely grated orange zest

Nut butter for serving, optional

Jam or marmalade for serving, optional

1. Preheat the oven to 425°F. Oil a baking sheet.

2. In a large bowl, combine the flour, bran, flaxseed, wheat germ, baking powder, baking soda, cinnamon, mace, brown sugar, dried fruit, and sweet potato. Stir in the yogurt, egg, oil, and orange zest and mix just until combined.

3. With lightly floured hands, knead the dough in the bowl for several minutes. When smooth, turn the dough onto a lightly floured work surface and pat to a thickness of ¼ inch. Using a cookie cutter or a glass, cut into 2½-inch rounds. Gather the scraps, gently pat into a ball, flatten again, and cut out more scones.

4. Place the scones on the prepared baking sheet about 1 inch apart. Bake for 8 to 10 minutes, or until golden brown. Serve warm with or without nut butter and jam or marmalade.

With the breakfast scones, have a grape cooler. Combine equal parts of chilled unsweetened concord grape juice and club soda and serve with a squeeze of lime.

CRAB-STUFFED PAPAYA

SERVES 4

Couscous

½ cup whole wheat couscous
½ teaspoon canola or safflower oil
⅛ cup chopped fresh flat-leaf parsley
1 teaspoon fresh lemon juice

Stuffed papayas

¾ cup diced celery
12 snowpeas
2 medium papayas (about ½ pound each), halved lengthwise and seeded
12 ounces diced cooked crabmeat
2 teaspoons fresh lemon juice
2 teaspoons curry powder
2 tablespoons finely sliced chives
2 tablespoons toasted sunflower seeds, see Note
Salt and freshly ground black pepper to taste

Garnishes

8 large leaves red leaf lettuce
4 orange wedges

1. To prepare the couscous, pour it into a medium-sized bowl and pour ½ cup boiling water over it. Cover and let stand for 5 minutes, or until tender. Fluff with a fork. Add the oil, parsley, and lemon juice and toss gently. Set aside.

2. To blanch the vegetables for the stuffed papayas: in a medium saucepan, bring 3 to 4 cups water to a boil. Place

the diced celery into a heatproof, waterproof strainer and immerse the celery in the boiling water for 1 minute. Lift the strainer and celery out of the water and cool under cold running water for 30 seconds. Drain well, transfer the celery to a small container, and set aside. Place the snowpeas into the strainer and repeat.

3. For the salad, use a small melon baller to make papaya balls, leaving just enough papaya in the shells so that the shells retain their shape. Set the shells aside and place the papaya balls into a medium mixing bowl. Add the crabmeat, blanched celery, lemon juice, curry powder, chives, sunflower seeds, and salt and pepper and toss to blend. Add the couscous and mix gently. Spoon equal portions of the mixture into the 4 papaya shells.

4. Place 2 red leaf lettuce leaves on each of 4 salad plates. Place a stuffed papaya half on each plate. Garnish with 3 snowpeas and an orange wedge and serve.

NOTE: To toast sunflower seeds, spread them in a small dry skillet and heat over medium-high heat for 30 to 45 seconds, shaking the pan until the seeds are fragrant and lightly browned. Immediately slide them from the pan to a plate to halt cooking.

SPINACH, MUSHROOM, AND EGG CHOPPED SALAD

3 hard-cooked omega-3 eggs, chilled
2 cups chopped fresh baby spinach leaves
¾ cup diced white mushrooms
Raspberry Vinaigrette (page 254)

1. To make the salad, discard 2 of the egg yolks or reserve for another use. Chop 1 egg yolk and all 3 egg whites and toss to mix.

2. In a large glass bowl, combine the baby spinach, mushrooms, and chopped eggs. Toss with ⅓ cup of the vinaigrette and serve immediately.

■ *Dinner*

ORANGE-GINGER CHICKEN WITH APRICOT-ALMOND COUSCOUS

....

SERVES 4

¾ cup fresh orange juice
3 tablespoons low-sodium soy sauce
3 teaspoons fresh minced ginger
4 boneless, skinless chicken breasts (4–5 ounces each)
2 tablespoons olive oil
20 green beans
8 baby carrots
1 medium red bell pepper, julienned
Apricot-Almond Couscous (page 298)
4 sprigs fresh flat-leaf parsley

1. In a shallow glass or ceramic bowl, mix together the orange juice, soy sauce, and ginger. Pour about half of the marinade into a storage container, cover, and refrigerate until needed. Leave enough marinade in the bowl to cover the chicken. Add the chicken breasts, turn several times to coat, cover, and refrigerate for at least 1 hour and up to 6 hours, turning several times.

2. In a nonstick sauté pan, heat the olive oil over medium heat. Lift the chicken breasts from the marinade and let the marinade drip back into the bowl and discard. Sauté the breasts for 2 to 3 minutes on each side or until golden brown. Spoon the reserved, unused marinade over the breasts. Cover, reduce the heat, and simmer for 5 to 8 minutes, or until the juices run clear when the meat is pierced with a fork or small sharp knife.

3. Meanwhile, in a steaming basket set over boiling water, steam the beans, carrots, and red pepper for 3 to 4 minutes or until fork tender.

4. Spoon the couscous onto plates and put a chicken breast on top of the couscous. Coat the chicken with pan juices and spoon the green beans, carrots, and peppers on the plate. Garnish with a sprig of parsley. Repeat to make 3 more servings.

APRICOT-ALMOND COUSCOUS

MAKES ABOUT 2 CUPS

½ cup diced dried apricots, diced
¼ cup orange juice
1¼ cups vegetable broth or water
1 cup whole wheat couscous
Pinch of cayenne pepper
⅓ cup toasted sliced almonds
1 tablespoon orange zest

1. In a small bowl, combine the apricots and orange juice and set aside for about 15 minutes to give the apricots time to plump.

2. In a saucepan, bring the vegetable broth to a boil over high heat.

3. In a heatproof mixing bowl, combine the couscous and cayenne pepper. Pour the boiling broth over the couscous and let it stand for 5 minutes. Fluff with a fork. Stir in apricots, any orange juice, the almonds, and zest. Serve immediately.

Enjoy with ½ cup nonfat frozen yogurt for dessert.

SESAME STIR-BRAISED KALE

SERVES 8

1 teaspoon sesame oil
4 garlic cloves, minced
1 tablespoon minced fresh ginger
2 to 3 tablespoons vegetable stock or water
2 bunches kale (about 1 pound), stemmed and chopped
1 teaspoon low-sodium soy sauce
1 tablespoon sesame seeds

1. In a large nonstick skillet, heat the oil over medium-high heat. Add the garlic and ginger and cook, stirring often, for about 1 minute, or until softened.

2. Add the stock and kale. Reduce the heat to low, cover, and cook for 2 to 3 minutes, or until the kale is wilted and tender. Drain excess liquid, if necessary. Add the soy sauce and toss gently. Cover and let stand for at least 20 minutes and up to 1 hour to allow the flavors to meld.

3. To serve, transfer the kale to a serving platter and sprinkle with sesame seeds. Toss and serve.

Nutrient Analyses

All of the *SuperFoods Rx* daily meal plans have been ana-lyzed by a registered nutritionist and measured utilizing the latest nutritional computer software and nutrient databases to ensure that all nutrient goals are met or exceeded on a daily basis. We measured all daily menus for a complete vitamin-mineral-protein-fat-calorie breakdown, with spe-cial emphasis on the intake of the fourteen SuperFoods, the foundation of the *SuperFoods Rx* program.

The *SuperFoods Rx* seven-day meal plan delivers the "super fourteen," along with the nutrients outlined in the Dietary Reference Intakes (DRIs) published by the Food and Nutrition Board of the Institute of Medicine. One nutri-ent that consistently is below the DRI is vitamin D. Supple-ments and/or sunshine are necessary to meet the recommended daily intake (DRI) for this vitamin.

Other nutrients that often fall below the published DRIs are calcium, molybdenum, biotin, and zinc. This reinforces the recommendation to take a daily multivitamin made on the SuperFoods Rx Lifestyle Pyramid.

It is not necessary to be a slave to the daily nutrient

goals especially when considering the carotenoids. For carotenoids, strive for your daily goal by averaging out over a week the intake of this important class of phytonutrients (e.g., the daily intake may be higher or lower than the daily amount mentioned, but over a week the average daily intake would be close to the recommended daily goal).

Please note that the SuperFoods Rx "super fourteen" nutrients are in boldface. For more information on them, see page 316.

DAY 1

	Unit	Value	Goal	% Goal
Calories		2448		
Protein	g	92		
Carbohydrate	g	332		
Fat (total)	g	91		
Cholesterol	mg	88		
Saturated	g	20		
Monounsaturated	g	38		
Polyunsaturated	g	21		
Linoleic (LA)	g	13.6	*	*
Linolenic (ALA)	g	3.7	see Omega-3s, page 321	surpass
EPA	g	0.0	1 g EPA/DHA	consider supplement
DHA	g	0.0		consider supplement
Omega-6:Omega-3	g	3.6:1	4:1 or less	surpass
Sodium	mg	2315		
Potassium	mg	7318		
Vitamins				
Vitamin A	RE	7084		

	Unit	Value	Goal	% Goal
Vitamin A	IU	70789		
Vitamin C	**mg**	**580**	**350**	**166%**
Vitamin D	IU	1.4		
Vitamin E	**mg**	**21.5**	**16**	**134%**
Thiamin	mg	2.0		
Riboflavin	mg	2.0		
Niacin	mg	19.7		
Pyridoxine (B$_6$)	mg	3.2		
Folate	**mcg**	**849**	**400**	**212%**
Vitamin B$_{12}$	mcg	3.1		
Biotin	mcg	24.9		
Pantothenic acid	mg	5.3		
Vitamin K	mcg	342		
Minerals				
Calcium	mg	1145		
Iron	mg	27.3		
Phosphorus	mg	1492		
Magnesium	mg	687		
Zinc	mg	14.1		
Copper	mg	3.1		
Manganese	mg	6		
Selenium	**mcg**	**41**	**70**	**59%**
Chromium	mcg	75		
Molybdenum	mcg	20		
Fiber	**g**	**53**	**see Fiber, page 321**	**surpass**
Carotenoids				
Lycopene	**mg**	**28.9**	**22**	**131%**
Lutein/Zeaxanthin	**mg**	**12.5**	**12**	**104%**
Alpha-carotene	**mg**	**3.0**	**2.4**	**125%**
Beta-carotene	**mg**	**24**	**6**	**400%**
Beta cryptoxanthin	**mg**	**1.5**	**1**	**151%**
Glutathione		**present**	**NA**	**NA**

	Unit	Value	Goal		% Goal
Resveratrol		present**	NA		NA
Polyphenols		present	NA		NA

■ *Percentage of Calories*

Protein   15%

Carbohydrates   53%

Fat   32%

*Amount depends on omega-3 intake per day

**Can achieve with 4 ounces purple grape juice or red wine

DAY 2

	Unit	Value	Goal	% Goal
Calories	g	2186		
Protein	g	99		
Carbohydrate	g	293		
Fat (total)	g	80		
Cholesterol	mg	191		
Saturated	g	13		
Monounsaturated	g	37		
Polyunsaturated	g	22		
Linoleic (LA)	g	13.8	**	**
Linolenic (ALA)	g	2.8	see Omega-3s, page 321	surpass
EPA	g	.4	see Omega-3s, page 321	surpass
DHA	g	1.3	see Omega-3s, page 321	surpass
Omega-6:Omega-3		3.1:1	4:1 or less	surpass
Sodium	mg	1916		
Potassium	mg	7259		

	Unit	Value	Goal	% Goal
Vitamins				
Vitamin A	RE	3607		
Vitamin C	**mg**	**620**	**350**	**177%**
Vitamin D	IU	103		
Vitamin E	**mg**	**18.2**	**16**	**114%**
Thiamin	mg	2.5		
Riboflavin	mg	2.9		
Niacin	mg	34.3		
Pyridoxine (B_6)	mg	3.4		
Folate	**mcg**	**684**	**400**	**171%**
Vitamin B_{12}	mcg	4.3		
Biotin	mcg	25.7		
Pantothenic Acid	mg	7		
Vitamin K	mcg	652		
Minerals				
Calcium	mg	957		
Iron	mg	27.3		
Phosphorus	mg	1694		
Magnesium	mg	642		
Zinc	mg	11.6		
Copper	mg	2.4		
Manganese	mg	9		
Selenium	**mcg**	**83**	**70–100**	**surpass**
Chromium	mcg	98		
Molybdenum	mcg	13.5		
Fiber		**63**	**see Fiber, page 321***	**surpass**
Carotenoids				
Lycopene	**mg**	**19.3**	**22**	**88%**
Lutein/Zeaxanthin	**mg**	**25.7**	**12**	**214%**
Alpha-Carotene	**mg**	**2.14**	**2.4**	**89%**
Beta-Carotene	**mg**	**12.85**	**6**	**107%**
Beta Cryptoxanthin	**mg**	**1.47**	**1**	**147%**

	Unit	Value	Goal	% Goal
Glutathione		present	NA	NA
Resveratrol		absent**	NA	NA
Polyphenols		present	NA	NA

▪ *Percentage of Calories*

Protein   17%

Carbohydrates   51%

Fat   31%

*Amount depends on omega-3 intake per day

**Can achieve with 4 ounces purple grape juice or red wine

DAY 3

	Unit	Value	Goal	% Goal
Calories	g	2040		
Protein	g	93		
Carbohydrate	g	265		
Fat (total)	g	74		
Cholesterol	mg	96		
Saturated	g	11		
Monounsaturated	g	35		
Polyunsaturated	g	18		
Linoleic (LA)	g	11.3	*	*
Linolenic (ALA)	g	3.7	see Omega-3s, page 321	surpass
EPA	g	.4	see Omega-3s, page 321	surpass
DHA	g	1.3	see Omega-3s, page 321	surpass
Omega-6:Omega-3		2:1	4:1 or less	surpass

	Unit	Value	Goal	% Goal
Sodium	mg	2192		
Potassium	mg	5131		
Vitamins				
Vitamin A	RE	4985		
Vitamin C	**mg**	**624**	350	178%
Vitamin D	IU	3.5		
Vitamin E	**mg**	**23.4**	16	146%
Thiamin	mg	1.7		
Riboflavin	mg	1.9		
Niacin	mg	23.2		
Pyridoxine (B$_6$)	mg	3.6		
Folate	**mcg**	**570**	400	143%
Vitamin B$_{12}$	mcg	4.4		
Biotin	mcg	20.4		
Pantothenic acid	mg	5.7		
Vitamin K	mcg	348		
Minerals				
Calcium	mg	1276		
Iron	mg	19.2		
Phosphorus	mg	1241		
Magnesium	mg	551		
Zinc	mg	9.8		
Copper	mg	2.5		
Manganese	mg	5.4		
Selenium	**mcg**	**63**	70–100	90%
Chromium	mcg	47		
Molybdenum	mcg	23		
Fiber		**42**	see Fiber, page 321*	surpass
Carotenoids				
Lycopene	**mg**	**23.9**	22	109%
Lutein/Zeaxanthin	**mg**	**12.1**	12	101%
Alpha-carotene	**mg**	**5.3**	2.4	219%

	Unit	Value	Goal	% Goal
Beta-carotene	mg	25.7	6	430%
Beta cryptoxanthin	mg	.87	1	87%
Glutathione		present	NA	NA
Resveratrol		absent **NA		NA
Polyphenols		present	NA	NA

■ *Percentage of Calories*

Protein   18%
Carbohydrates   51%
Fat   31%

*Amount depends on omega-3 intake per day
**Can achieve with 4 ounces purple grape juice or red wine

DAY 4

	Unit	Value	Goal	% Goal
Calories		2260		
Protein	g	128		
Carbohydrate	g	304		
Fat (total)	g	64		
Cholesterol	mg	474		
Saturated	g	12		
Monounsaturated	g	30		
Polyunsaturated	g	13		
Linoleic (LA)	g	7.2	*	*
Linolenic (ALA)	g	1.2	see Omega-3s, page 321	adult males — below adult females — surpass

	Unit	Value	Goal	% Goal
EPA	g	0.4	see Omega-3s, page 321	surpass
DHA	g	1.2	see Omega-3s, page 321	surpass
Omega-6:Omega-3	g	2.6:1	4:1 or less	surpass
Sodium	g	3407		
Potassium	g	6173		
Vitamins				
Vitamin A	RE	8673		
Vitamin C	mg	590	350	169%
Vitamin D	IU	40		
Vitamin E	mg	16.4	16	102%
Thiamin	mg	1.2		
Riboflavin	mg	2.2		
Niacin	mg	30		
Pyridoxine (B$_6$)	mg	3		
Folate	mcg	502	400	126%
Vitamin B$_{12}$	mcg	4.2		
Biotin	mcg	33		
Pantothenic acid	mg	7.4		
Vitamin K	mcg	552		
Minerals				
Calcium	mg	941		
Iron	mg	24.5		
Phosphorus	mg	1508		
Magnesium	mg	445		
Zinc	mg	10.6		
Copper	mg	1.7		
Manganese	mg	5.6		
Selenium	mcg	202	70–100	surpass
Chromium	mcg	57		
Molybdenum	mcg	38		

	Unit	Value	Goal	% Goal
Fiber		38	see Fiber, page 321*	surpass
Carotenoids				
Lycopene	mg	23.8	22	108%
Lutein/Zeaxanthin	mg	13.6	12	113%
Alpha-carotene	mg	8.8	2.4	366%
Beta-carotene	mg	22.7	6	378%
Beta cryptoxanthin	mg	1.5	1	150%
Glutathione		present	NA	NA
Resveratrol		present**	NA	NA
Polyphenols		present	NA	NA

▪ Percentage of Calories

Protein   22%
Carbohydrates   53%
Fat   25%

*Amount depends on omega-3 intake per day
**Can achieve with 4 ounces of purple grape juice or red wine

DAY 5

	Unit	Value	Goal	% Goal
Calories	g	1820		
Protein	g	127		
Carbohydrate	g	245		
Fat (total)	g	41		
Cholesterol	mg	340		
Saturated	g	7.8		
Monounsaturated	g	17.5		

	Unit	Value	Goal	% Goal
Polyunsaturated	g	8.8		
Linoleic (LA)	g	4.4	*	*
Linolenic (ALA)	g	2.8	see Omega-3s, page 321	surpass
EPA	g	.1	see Omega-3s, page 321	consider supplement
DHA	g	.5	see Omega-3s, page 321	consider supplement
Omega-6:Omega-3	g	1.3:1	4:1 or less	surpass
Sodium	mg	3094		
Potassium	mg	6184		
Vitamins				
Vitamin A	RE	4937		
Vitamin C	mg	419	350	120%
Vitamin D	IU	880		
Vitamin E	mg	16	16	100%
Thiamin	mg	1.8		
Riboflavin	mg	2		
Niacin	mg	30.3		
Pyridoxine (B6)	mg	3.8		
Folate	mcg	694	400	173%
Vitamin B12	mcg	4.2		
Biotin	mcg	26.8		
Pantothenic acid	mg	7		
Vitamin K	mcg	603		
Minerals				
Calcium	mg	1118		
Iron	mg	27.2		
Phosphorus	mg	1747		
Magnesium	mg	644		
Zinc	mg	13.9		
Copper	mg	2.2		
Manganese	mg	4.6		

	Unit	Value	Goal	% Goal
Selenium	mcg	138		**surpass**
Chromium	mcg	62		
Molybdenum	mcg	38.8		
Fiber		48	see Fiber, page 321	**surpass**
Carotenoids				
Lycopene	mg	26.6	22	121%
Lutein/Zeaxanthin	mg	20.8	12	173%
Alpha-carotene	mg	4.1	2.4	171%
Beta-carotene	mg	14	6	233%
Beta cryptoxanthin	mg	1.2	1	120%
Glutathione		present	NA	NA
Resveratrol		present **	NA	NA
Polyphenols		present	NA	NA

▪ *Percentage of Calories*

Protein   27%
Carbohydrates   53%
Fat   20%

*Amount depends on omega-3 intake per day

**Can achieve with 4 ounces purple grape juice or red wine

DAY 6

	Unit	Value	Goal	% Goal
Calories	g	1997		
Protein	g	103		
Carbohydrate	g	288		
Fat (total)	g	57		
Cholesterol	mg	221		

	Unit	Value	Goal	% Goal
Saturated	g	9.6		
Monounsaturated	g	20		
Polyunsaturated	g	21.8		
Linoleic (LA)	g	12.7	*	*
Linolenic (ALA)	g	6.2	see Omega-3s, page 321	surpass
EPA	g	trace	see Omega-3s, page 321	consider supplement
DHA	g	trace	see Omega-3s, page 321	consider supplement
Omega-6:Omega-3	g	2.1:1	4:1 or less	surpass
Sodium	mg	1803		
Potassium	mg	4635		
Vitamins				
Vitamin A	RE	7393		
Vitamin C	mg	355	350	101%
Vitamin D	IU	114		
Vitamin E	mg	15.8	16	98%
Thiamin	mg	1.9		
Riboflavin	mg	2		
Niacin	mg	23.4		
Pyridoxine (B_6)	mg	3.3		
Folate	mcg	587	400	147%
Vitamin B_{12}	mcg	1.2		
Biotin	mcg	31.8		
Pantothenic acid	mg	6.5		
Minerals				
Calcium	mg	1205		
Iron	mg	21.9		
Phosphorus	mg	1828		
Magnesium	mg	630		
Zinc	mg	12.6		
Copper	mg	2.6		

	Unit	Value	Goal	% Goal
Manganese	mg	8.7		
Selenium	mcg	110	70–100	surpass
Chromium	mcg	38		
Molybdenum	mcg	31.5		
Fiber		48	see Fiber, page 321	surpass
Carotenoids				
Lycopene	mg	35.9	22	163%
Lutein/Zeaxanthin	mg	11.8	12	98%
Alpha-carotene	mg	9.4	2.4	392%
Beta-carotene	mg	19.8	6	333%
Beta cryptoxanthin	mg	1.6	1	160%
Glutathione		present	NA	NA
Resveratrol		present **	NA	NA
Polyphenols		present	NA	NA

▪ *Percentage of Calories*

Protein   20%
Carbohydrates   55%
Fat   25%

*Amount depends on omega-3 intake per day
**Can achieve with 4 ounces purples grape juice or red wine

DAY 7

	Unit	Value	Goal	% Goal
Calories	g	1909		
Protein	g	110		
Carbohydrate	g	254		
Fat (total)	g	59		

	Unit	Value	Goal	% Goal
Cholesterol	mg	226		
Saturated	g	7.9		
Monounsaturated	g	17.9		
Polyunsaturated	g	21.6		
Linoleic (LA)	**g**	**16.3**	*	*
Linolenic (ALA)	**g**	**3.9**	see Omega-3s, page 321	surpass
EPA	**g**	trace	see Omega-3s, page 321	consider supplement
DHA	**g**	trace	see Omega-3s, page 321	consider supplement
Omega-6:Omega-3	**g**	**4.1:1**	4:1 or less	met
Sodium	mg	2047		
Potassium	mg	4123		
Vitamins				
Vitamin A	RE	3752		
Vitamin C	**mg**	**352**	350	100%
Vitamin D	IU	22		
Vitamin E	**mg**	**14.8**	16	87%
Thiamin	mg	.9		
Riboflavin	mg	1.3		
Niacin	mg	26		
Pyridoxine (B$_6$)	mg	1.9		
Folate	**mcg**	**384**	400	96%
Vitamin B$_{12}$	mcg	1.0		
Biotin	mcg	22.2		
Pantothenic acid	mg	5.2		
Vitamin K	mcg	614		
Minerals				
Calcium	mg	856		
Iron	mg	13.8		
Phosphorus	mg	1027		
Magnesium	mg	334		

	Unit	Value	Goal	% Goal
Zinc	mg	6.5		
Copper	mg	1.8		
Manganese	mg	5.3		
Selenium	mcg	1490	70–100	surpass
Chromium	mcg	44		
Molybdenum	mcg	0.9		
Fiber		36	see Fiber, page 321	surpass
Carotenoids				
Lycopene	mg	22	22	100%
Lutein/Zeaxanthin	mg	30.3	12	253%
Alpha-carotene	mg	3.6	2.4	150%
Beta-carotene	mg	16.4	6	273%
Beta cryptoxanthin	mg	.25	1	25%
Glutathione		present	NA	NA
Resveratrol		present**	NA	NA
Polyphenols		present	NA	NA

▪ Percentage of Calories

Protein 22%
Carbohydrates 51%
Fat 27%

*Amount depends on omega-3 intake per day

**Can achieve with 4 ounces purple grape juice or red wine

The Fourteen Super Nutrients

If you analyze all the most health-promoting, disease preventing, anti-aging, risk-factor-limiting diets in the world, fourteen nutrients consistently turn up. They are associated with reducing a wide range of chronic ailments. Countless studies demonstrate that the higher your level of these nutrients, the slower you age and the less chronic disease you suffer. Here is a list of the top fourteen Super Nutrients along with the foods that offer the richest sources of the "super fourteen."

If you have a bleeding or clotting problem, or are taking anticoagulants, consult your health care professional before adopting any of the recommendations listed in this appendix.

One: Vitamin C
Aim for at least 350 milligrams per day from a combination of the following foods:

1 large yellow bell pepper = 341 mg.

1 large red bell pepper = 312 mg.

1 common guava = 165 mg.

1 large green bell pepper = 132 mg.
1 cup fresh orange juice = 124 mg. (97 mg./cup from frozen concentrate)
1 cup fresh sliced strawberries = 97 mg.
1 cup fresh broccoli (chopped) = 79 mg.

Two: Folic Acid

Aim for 400 micrograms per day from a combination of the following foods:

1 cup cooked spinach = 263 mcg. folic acid (in food folic acid is called folate)
1 cup boiled kidney beans = 230 mcg.
1 cup boiled green soybeans = 200 mcg.
½ cup soy nuts = 177 mcg.
1 cup orange juice from frozen concentrate = 110 mcg.
4 cooked asparagus spears with ½ inch base = 89 mcg.
1 cup (frozen) chopped cooked broccoli = 103 mcg.

Three: Selenium

Aim for 70 to 100 micrograms per day from a combination of the following foods:

3 ounces cooked Pacific oysters = 131 mcg.
1 cup whole grain wheat flour = 85 mcg.
1 dried Brazil nut = 68 to 91 mcg.
½ can of Pacific sardines = 75 mcg.
3 ounces of canned white tuna = 56 mcg.
3 ounces cooked clams = 54 mcg.
6 farmed oysters = 54 mcg.
3 ounces roasted skinless turkey breast = 27 mcg.

Four: Vitamin E

Aim for at least 16 milligrams per day from a combination of the following foods:

2 tablespoons wheat germ oil = 41 mg. (total tocopherals)

2 tablespoons soybean oil = 2.6 mg.
2 tablespoons canola oil = 13.6 mg.
2 tablespoons peanut oil = 9.2 mg.
2 tablespoons flaxseed oil = 4.8 mg.
2 tablespoons olive oil = 4 mg.
1 ounce raw (23–24 whole kernels) almonds = 7.7 mg.
¼ cup hulled dry-roasted sunflower seeds = 6.8 mg.
2 tablespoons raw (untoasted) wheat germ = 5 mg.
1 medium orange bell pepper = 4.3 mg.
1 ounce hazelnuts (20–21 kernels) = 4.3 mg.
2 tablespoons peanut butter = 3.2 mg.
1 cup blueberries = 2.8 mg.

Five: Lycopene

Aim for 22 milligrams per day of this carotenoid from a combination of the following foods:

   1 cup tomato sauce (canned) = 37 mg.
   1 cup R.W. Knudsen Very Veggie vegetable cocktail from concentrate = 22 mg.
   1 cup tomato juice = 22 mg.
   1 watermelon wedge (¹⁄₁₆ of a melon 15 inches long, 7½ inches in diameter) = 13 mg.
   1 cup canned stewed tomatoes = 10.3 mg.
   1 tablespoon tomato paste = 4.6 mg.
   1 tablespoon ketchup = 2.9 mg.
   ½ pink grapefruit = 1.8 mg.

Keep in mind that tomato sources of lycopene are far more bioavailable from cooked vs. raw (unprocessed) tomato products.

The lycopene found in watermelon is very bioavailable. (To date we are not aware of any studies evaluating the absorption characteristics of other fruit sources of lycopene; presumably they are similar to watermelon.)

Six: Lutein/Zeaxanthin

Aim for 12 milligrams per day of this carotenoid from a combination of the following foods:

1 cup cooked kale (chopped) = 23.7 mg.
1 cup cooked spinach = 20.4 mg.
1 cup cooked collard greens (chopped) = 14.6 mg.
1 cup cooked turnip greens = 12.1 mg.
1 large sweet orange bell pepper = 9.2 mg.
1 cup cooked green peas = 4.2 mg.
1 cup cooked broccoli = 2.4 mg.

Seven: Alpha-carotene

Aim for 2.4 milligrams per day of this carotenoid from a combination of the following foods:

1 cup canned pumpkin = 11.7 mg.
1 cup cooked carrots (slices) = 6.6 mg.
10 raw medium baby carrots = 3.8 mg.
1 cup cooked butternut squash (cubes) = 2.3 mg.
1 large sweet orange bell pepper = .3 mg.
1 cup cooked collard greens (chopped) = .2 mg.

Eight: Beta-carotene

Aim for 6 milligrams per day of this carotenoid from a combination of the following foods:

1 cup cooked sweet potato = 23 mg.
1 cup canned pumpkin = 17 mg.
1 cup cooked carrots (slices) = 13 mg.
1 cup cooked spinach = 11.3 mg.
1 cup cooked chopped kale = 10.6 mg.
1 cup cooked butternut squash (cubes) = 9.4 mg.
1 cup cooked collard greens (chopped) = 9.2 mg.

Nine: Beta Cryptoxanthin

Aim for at least 1 milligram per day of this carotenoid from a combination of the following foods:

　　1 cup cooked butternut squash (cubes) = 6.4 mg.
　　1 cup cooked red bell pepper (strips) = 2.8 mg.
　　1 Japanese persimmon (2½ inches in diameter) = 2.4 mg.
　　1 cup mashed papaya = 1.8 mg.
　　1 large sweet red bell pepper (raw) = .8 mg.
　　1 cup fresh tangerine juice = .5 mg.
　　1 medium tangerine = .3 mg.

Ten: Glutathione

Optimum daily recommendation amounts are not yet known. Foods high in glutathione include:

　　asparagus
　　watermelon
　　avocado
　　walnuts
　　grapefruit
　　peanut butter
　　oatmeal
　　broccoli
　　oranges
　　spinach

Eleven: Resveratrol

Optimum daily recommendation amounts are not as yet known. Data suggests that this phytonutrient plays a role in preventing inflammation and cancer. It seems to have cardio-protective activity. Foods high in resveratrol include:

　　peanuts
　　purple grape skins
　　red wine

purple grape juice
cranberries/cranberry juice

Twelve: Fiber

The Food and Nutrition Board of the Institute of Medicine recently released their new dietary reference intakes for fiber. They are:

Females 19 to 50 years old: 25 grams; females 51 to 70 years old: 21 grams

Males 19 to 50 years old: 38 grams; males 51 to 70 years old: 30 grams

I feel these should be minimum goals, and if one's fiber intake is higher through the consumption of whole food and whole food products, so much the better.

Whole foods:

1 cup cooked black beans = 15 grams fiber

¼ cup dry pinto beans = 14 grams

1 cup cooked garbanzo beans = 13 grams

¼ cup dry lentils = 9 grams

1 cup fresh raspberries = 8 grams

Thirteen: Omega-3 Fatty Acids

The Food and Nutrition Board of the Institute of Medicine, the National Academies, recently set an adequate intake of 1.6 grams per day of plant-derived omega-3s (alpha linolenic acid, ALA) for adult men and 1.1 grams per day for adult women. They set a target amount of marine-derived omega-3s (EPA/DHA) of 160 milligrams per day for adult males and 110 milligrams per day for adult females.

I agree with the Food and Nutrition Board's daily ALA recommendations (they are great minimums; if you reach higher daily intake levels eating the *Superfoods Rx* way, that is okay). My personal daily recommendations for marine-derived omega-3s are 1 gram per day of EPA/DHA for adult males and .7 gram (700 milligrams) for adult females.

Aim for the above goals through a combination of the following foods:

EPA/DHA = primarily marine-derived omega-3 food sources

    3 ounces cooked Chinook (king) salmon = 1.5 grams
    3 ounces sockeye salmon = 1 gram
    3 ounces farmed Rainbow trout = 1 gram
    1 can sardines = .9 gram
    3 ounces canned white tuna in water = .7 gram

Consider using an EPA/DHA supplement on days when you do not eat marine sources of omega-3 fats.

Alpha linolenic acid (ALA)=plant-derived omega-3 food sources

Oils
1 tablespoon canola oil = 1.3 grams
1 tablespoon soybean oil = .7 gram
1 tablespoon walnut oil = 1.4 grams
1 tablespoon flaxseed oil = 7.3 grams

Green leafies
1 cup cooked spinach = .2 gram
1 cup cooked collard greens = .2 gram

Other foods
½ cup dry roasted soy nuts = 1.2 grams
1 tablespoon flaxseed = 2.2 grams
½ cup wheat germ = .5 gram
1 ounce (14 halves) English walnuts = 2.6 grams

1 omega-3 "vegetarian" hen egg = amounts vary; check
the carton

Fourteen: Polyphenols

Optimum daily recommended amounts for this class of phytonutrients has not yet been determined. The following foods and beverages have significant amounts of polyphenols:

Whole foods
berries
dates and figs
prunes
kale, spinach
parsley, dried parsley
apples with skin
citrus
grapes

Jams
The top three jams we tested for total polyphenol content:
  Trader Joe's Organic Blueberry Fruit Spread
  Knott's Pure Boysenberry Preserves
  Trader Joe's Organic Blackberry Fruit Spread

Beverages
green, black, or oolong tea
soymilk
100 percent fruit juices (berry, pomegranate, Concord
  grape, cherry, apple, citrus, prune)
The top three 100 percent fruit juices we tested for total polyphenol content were Odwalla C Monster, Trader Joe's 100 percent Unfiltered Concord Grape Juice, and R.W. Knudsen 100 percent Pomegranate Juice.

SuperFoods Rx
Shopping Lists

We all know that many of the foods available to us are not really health-promoting. Many fast foods, processed foods, and widely available packaged foods have too much fat and trans fat, sodium, sugar, and other nutrition negatives. It can be discouraging to shop. But there is some good news: there are many foods available in local markets that are good for you and also delicious. I checked thousands of foods in supermarkets and other popular chain stores across the country. Many foods didn't make the cut. The ones that did, I've listed here in a SuperFoods Rx Shopping List. (This isn't meant to be a definitive list and there are other healthy foods not listed here.) These foods meet my own SuperFood criteria: they're all fairly close to whole foods—which is to say, they're minimally, or "healthfully," processed and they're low or at least relatively low in sodium and fat (no trans fats) and high in fiber and nutrients. They're health-promoting choices that you'll be happy to have in the fridge or your pantry.

A few of these foods are available by mail order; a few are not widely available. In those cases, I've listed the e-mail or street address so you can contact them if you choose.

I was pleased to find that some chain markets—Whole Foods, Trader Joe's, and Costco, for example—carry some excellent food choices, and I list them here. I hope as we consumers demand more healthful foods, we'll be able to find these foods and more like them available everywhere.

Costco Wholesale Corp.	Trader Joe's Markets	Whole Foods Market
P.O. Box 34535	*www.traderjoes.com*	Austin, TX 78703
Seattle, WA	800-746-7857	*www.wholefoods.com*
98124-1535		
1-800-774-2678		

Applesauce

- Mott's Applesauce Natural Unsweetened
- Trader Joe's Unsweetened Applesauce

- Trader Joe's Chunky Spiced Apples— naturally sweetened with apple juice concentrate

Bottled Fruit/Vegetable Juices

- Ceres 100% fruit juices— multiple flavors www.ceresjuices.com
- Evolution Fresh juices— multiple flavors www.evolutionfresh.com
- Hain Pure Foods Veggie Juice
- Hain Pure Foods Carrot Juice
- Kedem Concord Grape Juice 100% Pure Juice

- Kirkland Cranberry-Grape 100% Juice Blend (Costco)
- Kirkland Cranberry-Raspberry 100% Juice Blend (Costco)
- Kirkland 100% Juice (no sugar added—no artificial flavors or colors)—multiple flavors (Costco)

Bottled Fruit/Vegetable Juices (*continued*)

- Kirkland 100% Multi-Vitamin Juice Blast—multiple flavors (Costco)
- Kirkland 100% Grape Juice, Newman's Own (Costco)
- L & A Black Cherry Juice
    Langer Juice Company, Inc. Industry, CA 91745
- Lakewood 100% Fruit Juices Apricot
- Lakewood 100% Fruit Juices Prune Juice
- Lakewood 100% Fruit Juices Peach
- Lakewood 100% Fruit Juices Pure Blueberry
- Lakewood 100% Fruit Juices Pure Black Cherry—includes puree (fiber) from the fruit, fresh pressed
    Lakewood Products
    Miami, FL 33242-0708
- Campbell Low Sodium V8 100% Vegetable Juice
- Martinelli's Certified Organic Apple Juice
- Martinelli's Certified Organic Apple Juice—Unfiltered Apple Juice
- Minute Maid 100% Pure Apple Juice
- Minute Maid 100% Pure Grape Juice
- Minute Maid 100% Pure Squeezed Orange Juice
- Minute Maid High Pulp Premium Calcium + D Home Squeezed Style 100% Pure Squeezed Orange Juice
- Mountain Sun Pure Cranberry—unsweetened www.mountainsun.com All 100% pure unsweetened cranberry juice is tart, so mix it with another, sweeter 100% pure fruit juice
- Naked Juice—multiple flavors (my favorite is Naked Superfood Food-Juice Power C) www.nakedjuice.com
- 100% Nantucket Nectors Premium Orange Juice www.juiceguys.com
- Ocean Spray Premium 100% Juice—no sugar added— multiple flavors
- Odwalla Fruit Juice Drink— multiple flavors (my favorites are C Monster and Tangerine 100% Pure Squeezed Juice) www.odwalla.com
- Organic Sir Real State of the Art Fruit Juices www.sirreal.com
- R.W. Knudsen Family Cherry Cider

- R.W. Knudsen Family Just Blueberry
- R.W. Knudsen Family Just Boysenberry
- R.W. Knudsen Family Just Concord
- R.W. Knudsen Family Organic Apple
- R.W. Knudsen Family Organic Cranberry Juice
- R.W. Knudsen Family Organic Cranberry Nectar
- R.W. Knudsen Family Organic Prune Juice
- R.W. Knudsen Family Organic Tomato Juice
- R.W. Knudsen Family Peach Nectar
- R.W. Knudsen Family Pineapple Nectar
- R.W. Knudsen Family Pomegranate Juice
- R.W. Knudsen Family Very Veggie Vegetable Cocktail (22 mg. of lycopene per serving)
- Samantha Body Zoomers— multiple flavors (my favorite is The Strawberry Desperately Seeking C Antioxidant Fruit Drink)
  www.freshsamantha.com

- Sunsweet not from concentrate Prune Juice with Pulp
- Sunsweet Prune Juice Plus (with lutein)
- Trader Joe's 100% Unfiltered Concord Grape Juice
- Trader Joe's All Natural Unfiltered Concord Grape Juice
- Tropicana 100% Pure Orange Juice
- Tropicana 100% Red Ruby Grapefruit Juice
- Walnut Acres Certified Organic Concord Grape Juice
- Walnut Acres Certified Organic Cherry Juice
- Walnut Acres Certified Organic Wild Cranberry
  www.walnutacres.com
- Welch's 100% Grape Juice
  Many markets private-label their fruit juice—look for 100% juice, no added sugar, preservatives, or artificial flavors/colors.

Bottled Sweet Peppers

- Mancini Sweet Roasted Peppers
  www.mancinifoods.com

Bread

- Arnold's 100% Whole Wheat 9 Grain Bread
- The Baker 9-Grain Whole Wheat
- The Baker Honey Cinnamon Raisin
- The Baker Seeded Whole Wheat
- The Baker Whole Grain Rye
- The Baker 7-Grain Sourdough Whole Wheat
  www.the-baker.com
- Food for Life Sprouted Hot Dog Buns
- Food for Life Sprouted Wheat Burger Buns
  1-800-797-5090
  www.food4life.com
- Milton's Healthy Multi-Grain Bread
  www.miltonsbaking.com
- Natural Ovens Bakery Breads
  1-800-558-3535
  www.naturalovens.com/Products/bakery/For the Health Conscious
  They will ship products to your home.
- The Original 100% Flourless Sprouted Grain Bread Ezekiel 4:7 Low Sodium—organic grains
- The Original Bran for Life Bread—organic grains
  www.foodforlife.com
- Pepperidge Farms Natural Whole Grain 9 Grain
- Premium Sara Lee 100% Whole Wheat Sliced Bread
- Roman Meal California 100% Whole Wheat Bread
- Rudi's Organic Bakery Honey Sweet Whole Wheat Bread
  www.rudisbakery.com
- Vogel's Soy & Flaxseed Bread
- Vogel's Whole Wheat & Honey Bread—3rd party organic certified flour
  www.vogelsbread.com

Canned Albacore Tuna

- Bumble Bee Solid White Albacore Tuna in water
- Chicken of the Sea Low Sodium Chunk White Albacore Tuna
- Chicken of the Sea Chunk Light Tuna in canola oil

Both Bumble Bee and Chicken of the Sea have no-added-salt choices

- Kirkland Solid White Albacore Tuna packed in water (Costco)
- Star Kist Solid White Albacore Tuna in water
- Trader Joe's White Solid Albacore Tuna
- Trader Joe's White Solid Albacore Tuna—reduced sodium

Canned Clams and Crab

- Sea Watch Chopped Sea Clams
  SWI
  8978 Glebe Park Drive
  Easton, MD 21601-7004

- Trader Joe's Crab Meat
- Trader Joe's Chopped Sea Clams

Canned Salmon

- Bumble Bee Alaska Pink Salmon
- Bumble Bee Alaska Sockeye Red Salmon
- Chicken of the Sea Pink Salmon— traditional
  Includes the skin and bones—the healthiest way to eat salmon, as the content of omega-3 fats and calcium is higher than in the skin-and-bone-removed canned salmon.

- Libby's Red Salmon—packed from fresh Alaskan Sockeye
- Trader Joe's Alaska Pink Salmon
- Trader Joe's Red Salmon
  All canned salmon is Alaskan wild salmon. We recommend that you avoid Atlantic Salmon, as it is farm-raised.

Canned Trout

- Appel Fillets of Smoked Trout
  www.appel-feinkost.de

Canned Beans

- Eden Organic Black Beans
- Eden Organic Chili Beans
- Eden Organic Garbanzo
  Beans
- Eden Organic Kidney Beans
- Eden Organic Lentils
- Henry's Marketplace
  Cannellini Beans
- Henry's Marketplace Dark
  Red Kidney Beans
- Henry's Marketplace
  Garbanzo Beans
- Henry's Marketplace
  Pinto Beans
- Henry's Marketplace
  Refried Black Beans
    Henry's Marketplace
    El Cajon, CA 92020
    This is a San Diego chain
    of markets.

- Trader Joe's Bean Medley
- Trader Joe's Black Beans
- Trader Joe's Cannellini Beans
- Trader Joe's Organic Baked
  Beans
- Trader Joe's Organic Garbanzo
  Beans
- Trader Joe's Organic Kidney
  Beans
- Trader Joe's Organic Pinto
  Beans
- Trader Joe's Refried Black
  Beans with Jalapeño Peppers
- Westbrae Garbanzo Beans
- Westbrae Great Northern
  Beans
- Westbrae Kidney Beans
- Westbrae Natural Vegetarian
  Organic Black Beans
- Westbrae Organic Lentils

Canned Chili

- Health Valley 99% Fat-Free
  Vegetarian Chili—Spicy
  Black Bean

- Health Valley Mild Black Bean
- Health Valley Vegetarian Lentil
  1-800-434-4246

Canned Evaporated Milk

- Nestlé Carnation Evaporated Fat Free Milk—Vitamins A and D
  added

Canned Fruit

- Dole Pineapple Chunks—
  in own juice
- Trader Joe's All Natural 100%
  Hawaiian Pineapple Chunks—
  in unsweetened pineapple juice

Canned Olives

- Lindsay Large Pitted Olives

Canned Pumpkin

- Libby's 100% Pure Pumpkin

Canned Tomato Products

- Classico Di Napoli Spicy
  Red Pepper Pasta Sauce
- Classico Di Napoli Tomato
  & Basil Pasta Sauce
    1-888-337-2420
    www.classico.com
- Colavita Marinara 100%
  Natural
    www.colavita.com
- Emeril's Roasted Red
  Pepper Pasta Sauce
    www.bgfoods.com
    www.emerils.com
- Henry's Marketplace
  Low-Sodium Pasta Sauce
- Hunt's Tomato Sauce—
  no salt added
- Hunt's Tomato Paste
- Hunt's Tomato Paste—
  no salt added
- Muir Glen Chunky Tomato
  Sauce
- Muir Glen Crushed Tomato
  with Basil
- Muir Glen Organic Chunky
  Tomato and Herb Pasta Sauce
- Muir Glen Organic Ground
  Peeled Tomatoes
- Muir Glen Organic Whole
  Peeled Tomatoes
- Muir Glen Tomato Paste
- Muir Glen Tomato Puree
- Muir Glen Whole Peeled
  Tomatoes
- S & W Petite-Cut Diced
  Tomatoes in Rich, Thick Juice
- Walnut Acres Organic Zesty
  Basil Pasta Sauce
    www.walnutacres.com

Cereal

- Alpen Swiss Style Cereal—Original Alpen Naturally Delicious Swiss Style Cereal—Low fat
- Alpen Swiss Style Cereal—No Added Sugar or Salt Alpen Naturally Delicious Swiss Style Cereal—Low fat
- Arrowhead Mills Spelt Flakes—Organic
- Back to Nature Ultra Flax
- Back to Nature Granola
  www.organicmilling.com
- Barbara's Bakery Grain Shop High Fiber Cereal
- Barbara's Bite Size Shredded Oats Crunch Wholegrain
  www.barbarasbakery.com
- Bob's Red Mill Natural Raw Wheat Germ
- Bob's Red Mill Whole Ground Flaxseed Meal
- Breadshop's Granola Crunchy Oatbran with Almonds & Raisins
- Breadshop's Granola Cranberry Crunch Muesli
  www.hain-celestial.com/bread.html
- Chappaqua Crunch Simply Granola with Raspberries—multiple additional flavors
  1-800-488-4602
- Cheerios
- Familia No Added Sugar Swiss Muesli
- Familia Low Fat Granola
- Familia Original Recipe Muesli
- Health Valley Organic Oat Bran Flakes
- Health Valley Organic Amaranth Flakes
- Health Valley Organic Oat Bran O's
- Health Valley Organic Golden Flax Cereal
- Health Valley Soy Flakes Cereal—Original
- Health Valley Soy Original Cereal O's
- Healthy Fiber Multigrain Flakes
- Oat Bran O's Cereal
- Organic Fiber 7 Multigrain Flakes
- Honey Crunch Organic—The Baker Muesli
- Rich Food Enriched Bran Flakes
  www.supervalue
  storebrands.com
- Kashi Go Lean Protein/High Fiber Cereal & Snack
- Kashi Go Lean Seven Whole Grains & Sesame

- Kashi Go Lean Good Friends
- Kellogg's Complete Oat Bran Flakes
- Kellogg's Complete Wheat Bran Flakes
- Nature's Path Heritage O's
- Nature's Path Organic Blueberry Almond Muesli
- Nature's Path Organic Flax Plus Multibran Cereal
- Nature's Path Organic Heritage Muesli with Raspberries & Hazelnuts
- Nature's Path Organic Heritage Multigrain Cereal
- Nature's Path Organic Multigrain Oatbran Cereal
- Nature's Path Organic Optimum Power Breakfast Cereal-Flax-Soy-Blueberry www.naturespath.com
- Organic Weetabix Whole Grain Wheat Cereal

- Post Grape Nut Flakes
- Post The Original Shredded Wheat 'N Bran
- Post The Original Shredded Wheat
- Post The Original Spoon Size Shredded Wheat
- Stoneburhr Untoasted Wheat Germ
    1-206-938-3487
- Kretschmer Toasted Wheat Germ
- Trader Joe's Organic Golden Flax Cereal
- Ultra Omega Balance www.naturalovens.com/ Products/bakery/For the Health Conscious They will ship products to your home.
- Uncle Sam Cereal—Toasted Whole Grain Wheat Flakes with Crispy Whole Flaxseed

Cooked Cereal

- Hodgson Mill Oat Bran All Natural Hot Cereal
- Hot Rolled Wheat Cereal
- John McCann's Steel Cut Irish Oatmeal
- McCann's Imported Quick Cooking Irish Oatmeal
- McCann's Instant Irish Oatmeal—regular

- Mother's 100% Natural Rolled Oats
- Mother's 100% Natural Wholegrain Barley
- Quick1-Minute Quaker Oats
- Quaker Instant Oatmeal— Regular Flavor
- Old Fashioned Quaker Oats 100% Whole Grain

Cooked Cereal (*continued*)

- Quaker Oat Bran Hot Cereal
- The Silver Palate Thick & Rough Oatmeal
- Stone-Buhr Cracked Wheat Cereal
- Stone-Buhr 4 Grain Cereal Mate
  Stone-Buhr Cereals
  1-206-938-3487
- Wheatena Toasted Wheat Cereal

Chips and Snack Food

- Abuelita Stone Ground White Corn Tortilla Chips
  S & K Industries, Inc.
  Manassas Park, VA 20111
- Certified Organic Chips by Good Health—multiple flavors
  www.e-goodhealth.com
- Dirty's All Natural Potato Chips—multiple flavors (my favorite is honey mustard)
  www.dirtys.co
- Eat Smart All Natural Snacks—Snyder's of Hanover Veggie Crisps
  www.snydersofhanover.com
- Garden of Eatin All Natural Tortilla Chips/Black Bean—Sesame Blues
  1-800-434-4246
- GeniSoy Soy Crisps Apple Cinnamon Crunch
- GeniSoy Soy Crisps Roasted Garlic and Onion
  www.genisoy.com
- Glennys Low Fat Soy Crisps (Barbeque, Onion & Garlic, Lightly Salted, and Salt & Pepper flavors are recommended)
  www.glennys.com
- Guiltless Gourmet Baked Mucho Nacho Tortilla Chips—made with organic yellow corn
- Guiltless Gourmet Baked Spicy Black Bean on Blue Corn Tortilla Chips—made with organic blue corn
  www.guiltlessgourmet.com
- Just Veggies All Natural Snack Food
- Just Cherries
- Just Raspberries
- Just Blueberries
- Just Pineapple
- Just Carrots
- Just Corn
  www.justtomatoes.com
  The above foods are dehydrated.

- Kettle Tortilla Chips Five Grain Yellow Corn Certified 100% Organic
- Kettle Tortilla Chips Salsa and Mesquite Six Grain Yellow Corn
- Kettle Tortilla Chips Sesame, Rye and Caraway Certified 100% Organic
- Natural Lay's Thick Cut Country BBQ
- Natural Ruffles Sea Salted Reduced Fat Potato Chips
- Natural Tostitos Yellow Corn—made with organic corn
- Organic Just Soy Nuts
  www.justtomatoes.com

- Pennysticks Brand Oat Bran Honey Mustard Pretzel Nuggets with Soy Protein
  www.benzels.com
- Terra-A Delicious Potpourri of Exotic Vegetables–Chips
  www.terrachips.com
- Trader Joe's Roasted & Salted Soy Nuts
- Trader Joe's Dry Roasted Edamame—lightly salted
- Whole Foods Market 365 Organic—Organic Tortilla Chips Blue Corn—lightly salted
  www.wholefoods.com
- Wild Rice Snack Chips
  www.frwr.com

Cookies

- Health Valley Fat Free Apple Spice Cookies
- Health Valley Carob Chip Cookies
- Health Valley Oatmeal Raisin Cookies

- Health Valley Chocolate Raspberry Cookies
  1-800-558-3535
  www.naturalovens.com
  They will ship products to your home.
- Pistachio Almond Thins
  Delices de Bretagne
  258 Boulevard Lebeau
  Ville St. Laurent, QC
  Canada H4N 1R4

Crackers

- Ak-mak 100% Whole Wheat Stone Ground Sesame Cracker
    Ak-mak Bakeries
    89 Academy Ave.
    Sanger, CA 93657-2104
    1-559-875-5511
- Health Valley Low Fat French Onion Crackers
- Health Valley No Salt Added Bruschetta Vegetable Crackers
- Health Valley Low Fat Garden Herb Crackers
- Health Valley Cracked Pepper Crackers
- Health Valley Sesame Crackers
- Health Valley Stoned Wheat Crackers
- Health Valley Original Oat Bran Graham Crackers
- Health Valley Original Oat Crackers
- Health Valley Original Amaranth Graham Crackers
- Kavli All Natural Five Grain Crispbread—multiple additional flavors
- Kashi TLC Tasty Little Crackers
- Trader Joe's Rye Mini Toasts
- Wasa Fiber Rye Crispbread
- Wasa Oats Crispbread

Dips

- Athenos Mediterranean Spreads Hummus—Roasted Red Pepper
- Athenos Mediterranean Spreads Hummus—Greek Style
- Athenos Mediterranean Spreads Hummus—Original
    1-800-343-1976
- Tribe of Two Sheiks Classic Hummus
- Tribe of Two Sheiks Classic Hummus with Forty Spices
- Tribe of Two Sheiks Classic Hummus with Sweet Roasted Red Peppers
- Tribe of Two Sheiks Classic Hummus with Roasted Garlic
    www.twosheiks.com

Dried Fruit

- Elizabeth's Natural
  Cranberries
  1-631-243-1626
- Hadley Pitted Deglet Noor
  Dates (grown in California)
  Packed by: Hadley Date
  Gardens,
  Thermal, CA 92274
- Maiani Kirkland Pitted
  Dried Plums—Sweet Pitted
  Prunes (Costco)
- Melissa's Organic Produce:
  Dried Mango (just
  mango—from Mexico)
  Dried Thompson Seedless
  Grapes (just raisins—
  from U.S.A.)
  Dried Flame Seedless
  Grapes (just raisins—
  from U.S.A.)
  Dried Papaya (just
  papaya—from Sri Lanka)
  Dried Bing Cherries (plus
  organic cane sugar—
  from U.S.A.)
  Dried Blueberries (plus
  organic cane sugar—
  from U.S.A.)
  Dried Cranberries (plus
  organic cane sugar/canola
  oil—from U.S.A.)
  Dried Persimmons (just
  persimmons—from U.S.A.)

Dried Tomato (just tomato—
  from U.S.A.)
Pine Nuts (just pine nuts)
Roasted Soy Nuts (no salt—
  from U.S.A.)
www.melissas.com
- Ocean Spray Craisins Original
  Sweetened Dried Cranberries
- Pavich Organic Raisins
- Sun-Maid Raisins
- Sunsweet Our Premium
  Prunes—no preservatives
- Sunsweet California Grown
  Pitted Dates
- Sunview Certified Organically
  Grown Raisins
- Sunview Green Seedless
  Raisins
- Sunview Red Seedless
  Raisins
  www.sunviewmarketing.com
- Trader Joe's Bing Cherries
- Trader Joe's Black Mission
  Figs
- Trader Joe's Blenheim Variety
  Unsulfured Apricots
- Trader Joe's California Organic
  Thompson Seedless Raisins
- Trader Joe's California
  Thompson Seedless Raisins
- Trader Joe's Dried Berry
  Medley

Dried Fruit *(continued)*

- Trader Joe's Dried Blueberries
- Trader Joe's Dried Cranberries
- Trader Joe's Dried Organic Cranberries
- Trader Joe's Dried Wild Blueberries
- ✓ Trader Joe's Extra Large High Moisture Prunes
- Trader Joe's French Variety Non Sorbate Pitted Prunes
- Trader Joe's Imported Organic Apricots
- Trader Joe's Marionberries
- Trader Joe's Pitted Prunes
- Trader Joe's Pitted Tart Montgomery Cherries
- Trader Joe's Rainier Cherries—unsulfured
- Trader Joe's ShoEi California Grown Pitted Prunes— no preservatives
- Trader Joe's Non-Sorbate Pitted Prunes
- Trader Joe's Organic Imported Apricots

Eggs

- Cage Free Hens Giving Nature Organic Eggs
  www.givingnaturefoods. com
- Deb El Just Whites All Natural 100% Dried Egg Whites
  Distribution by: Deb-El Foods Corp.
  2 Papetti Plaza
  Elizabeth, NJ 07206
- Egg Beaters 99% Real Eggs
- Farm Fresh Egg Land's Best Grade A Eggs—Large All Natural Vegetarian Fed
  www.eggland.com
- Gold Circle Farms All Natural, Vegetarian Feed Extra Nutritious—2 Eggs provide 300 mg DHA Omega-3 Eggs
  1-888-599-4DHA
  www. goldcirclefarms.com
- Horizon Organic Extra Large Brown Eggs "Cage Free"
- Organic Omega-3 Eggs
  www.ChinoValleyRanchers. com
- Kirkland Egg Starts (Costco)

Enchilada Sauce
- Hatch Select Enchilada Sauce
    Hatch Chili Co.
    P.O. Box 752
    Deming, NM 88031

Energy Bars
- The Bagel B. B. Bakery
    Premium Quality Energy Bars
    The Bagel Brothers
- DBA: B. B. Bakery and
    Distributing, Inc.
    Costa Mesa, CA 92627
- Clif Bars (Cranberry Apple
    Cherry and Carrot Cake
    are recommended)
    www.clifbar.com
- Health Valley Fat Free
    Blueberry Granola Bars
- Health Valley Moist and
    Chewy Granola Bars
- Kashi Go Lean Vanilla Spice
    Cake
- Kashi Go Lean Oatmeal
    Raisin Cookie
- Power Bar (Chocolate,
    Vanilla Crisp, and Peanut
    Butter are recommended)
    www.powerbar.com

Frozen Beef
- Smart Meat A Whole New Grade of Beef—New York Strip Beef
    Steak
    —28% less cholesterol than skinless chicken breast
    —63% less total fat compared to USDA data
    GFI Premium Foods
    2815 Blaisdell Ave. South
    Minneapolis, MN 55408

Frozen Buffalo
- Great Range Brand Ground Buffalo
    Rocky Mountain Natural Meats, Inc.
    Denver, CO 80229

Frozen Burgers

Salmon Burgers

- Omega Foods Wild Salmon Burgers
  www.omegafoods.net

Turkey Burgers

- Pilgrims Pride Turkey Burgers—all white meat
  www.pilgrimspride.com

Veggie Beef Burgers

- Boca Meatless Burgers—original vegan
  www.bocaburger.com
- Lightlife Smart Ground Taco & Burrito (veggie burger)
  1-800-769-3279
  www.lightlife.com
- Veggie Ground Beef
  1-800-667-9837
  www.yvesveggie.com

Veggie Chicken Burgers

- Yves Veggie Cuisine Veggie Chick'n Burger
  1-800-667-9837
  www.yvesveggie.com

Veggie Burgers

- Dr. Praeger's Veggie Royale All Natural California Veggie Burgers
  www.drpraegers.com
- Gardenburger Veggie Medley
  www.gardenburger.com
- Whole Foods Market 365 Meat Free Gourmet Burger
- Vegan Burger
- Meat Free Garlic Burger
  www.wholefoods.com

Frozen Desserts

- Certified Organic Natural Choice Organic Sorbet (Blueberry and Strawberry-Kiwi are recommended)
  Natural Choice Foods, Inc.
  Oxnard, CA 93030
- Dreyer's Whole Fruit Sorbet (sold as Edy's Whole Fruit Sorbet east of the Rockies)
- Häagen-Dazs Fruit Sorbets—multiple flavors (my favorites are Mango and Raspberry)
- Häagen-Dazs Frozen Yogurt—multiple flavors
- Stonyfield Farm Non-Fat Frozen Yogurt—multiple flavors

Frozen Fish

- Cox's Delux Pink Shrimp
  Cox's Wholesale Seafood
  Tampa, FL 33684
- High Liner Atlantic Cod Fillets
  www.highlinerfoods.com
- Wild Alaskan Salmon
  www.oceanbeauty.com

Frozen Fruit Bars

- Dole Fruit Juice Quiescently Frozen Juice Bars
- Dole Fruit & Juice Frozen Bars Recommended flavors: Strawberry, Grape, Raspberry.
- Tropicana Premium Frozen Juice Bars (made with chunks of fruit) Recommended flavor: Strawberry.
- Dreyers Whole Fruit Bars (sold as Edy's Fruit Bars east of the Rockies) Recommended flavors: Strawberry, Lemonade, Lime, Wild Berry.
  www.dreyers.com
- Welch's Concord Grape Juice Bars

Frozen Fruit/Vegetables

- Cascadian Farms Organic Blackberries
- Cascadian Farms Organic Chinese Stir Fry
- Cascadian Farms Organic Garden Blend Premium Vegetables
- Cascadian Farms Organic Harvest Berries
- Cascadian Farms Organic Red Raspberries
- Cascadian Farms Organic Shelled Edamame
- Cascadian Farms Organic Strawberries
- Cascadian Farms Organic Sweet Cherries
  1-800-624-4123
  www.cfarm.com
- Flav. R. Pac Triple Berry Blend (blueberries/Marionberries/ raspberries)
  www.norpac.com
- Nutri Verde No Salt Added Vegetables—broccoli & cauliflower florets with slices of carrots, zucchini & yellow squash
  1-800-491-2665
  www.nutriverde.com

Frozen Fruit/Vegetables (*continued*)

- Pure Nature Organic Fruits & Vegetables—multiple choices www.purenatureorganics. com
- Trader Joe's Frozen Fruit/ Frozen Vegetables—multiple varieties
- Triple Berry Blend www.norpac.com Stores such as Trader Joe's and Whole Food Marketplace have their own brands of healthy frozen 100% vegetables and fruits.

Frozen Juices

Any brand that is 100% frozen concentrated juice with no added sugar and artificial colors.

Frozen Taquitos

- Whole Foods Market 365 Chicken Taquitos www.wholefoods.com

Frozen Waffles

- Kashi Go Lean Original 6 All Natural Frozen Waffles
- Lifestream Made with Organic Grains Flax Plus Toaster Waffles
- Lifestream Made with Organic Grains 8 Grain Seasame www.naturepath.com
- Van's All Natural Wheat Free Original Gourmet Waffles

Herbs/Spices/Flavor Enhancers

- It's Delish Garlic Granulated
- It's Delish Basil
- It's Delish Parsley Flakes
- It's Delish Oregano www.itsdelish.com
- McCormick Granulated Onion

Honey

- Gourmet Honey Store
  Buckwheat Honey
  www.gourmethoneystore.
  com
- Rita Miller's Select Honey
  Premium Gourmet Quality—
  Buckwheat
  www.millershoney.com

- Topanga Quality Honey—
  Buckwheat
  Bennet's Honey Farm
  Piru, CA 93040
  1-805-521-1375

Jams

- Knott's Berry Farm Pure
  Boysenberry Preserves
- Knott's Berry Farm Bing
  Cherry Pure Preserves
- Knott's Berry Farm 100%
  Fruit (Spreadable Fruit)—
  multiple flavors
- Sorrell Ridge Premium
  100% Fruit Wild Blueberry
  Spreadable Fruit
- Trader Joe's Organic
  Blueberry Fruit Spread
- Trader Joe's Organic
  Blackberry Fruit Spread
- Trader Joe's Organic
  Strawberry Fruit Spread

- Trader Joe's Organic Morello
  Cherry Fruit Spread
- Trader Joe's Blueberry
  Preserves made with Fresh
  Blueberries
- Trader Joe's Boysenberry
  Preserves made with Fresh
  Boysenberries
- Trader Joe's Strawberry
  Preserves made with Fresh
  Strawberries

Stores such as Whole Foods
Marketplace have their own
brand of spreadable fruit jams.

Ketchup

- Heinz Tomato Ketchup
- Muir Glen Organic Tomato
  Ketchup

- Trader Joe's Organic
  Ketchup

Margarine

- Smart Balance No-Trans Fatty Acids Buttery Spread

Nuts and Seeds

- Anne's Unsalted Dry Roasted Peanuts
    Only ingredient is dry-roasted blanched peanuts.
    Anne's House of Nuts Inc.
    Jessup, MD 20794
- David Roasted & Salted Sunflower Seeds
    ConAgra Foods
    7700 Frances Ave. South, Suite 200
    Edna, MN 56485
- Elizabeth's Natural Filberts
- Elizabeth's Natural Health Mix, Pecans
- Elizabeth's Natural No Salt Roasted in Shell Pumpkin Seeds
- Elizabeth's Natural Pepitas (Raw Shelled Pumpkin Seeds)
- Elizabeth's Natural Raw Almonds
- Elizabeth's Natural Raw Cashews
- Elizabeth's Natural Raw Hulled Sunflower Seeds
- Elizabeth's Natural Raw Mixed Nuts
- Elizabeth's Natural Super Energy Mix
- Elizabeth's Natural Walnuts
    1-631-243-1626
- Hoody's Classic Roast Peanuts—original nut house brand
    Original Nut House Brands
    11B Leigh Fisher
    El Paso, TX 79906
    The only ingredient is roasted-in-the-shell peanuts. This means the brown skin of the peanut is present, and this is where the polyphenols and resveratrol are concentrated.
- Kirkland Almonds—U.S. #1 Supreme Whole (Costco)
- Kirkland Pecan Halves (Costco)
- Kirkland Pine Nuts—raw pine nuts that's all (Costco)
- Kirkland Walnuts—preservative free shelled (Costco)

- Old Tyme Roasted Peanuts—
  no salt or oil
    These are in the shell,
    which means the brown
    skin of the peanut is
    present and this is where
    the polyphenols and
    resveratrol are
    concentrated.
    www.oldetymefoods.com
- Planters Salted Peanuts
- Trader Joe's California
  Premium Walnut Halves
- Trader Joe's California
  Walnut Halves & Pieces
- Trader Joe's Dry Roasted
  & Unsalted Almonds
- Trader Joe's Dry Roasted
  & Unsalted Pistachio
  Nutmeats
- Trader Joe's Dry Roasted &
  Unsalted Pistachios (in shell)
- Trader Joe's Fancy Dry
  Roasted Mixed Nuts
- Trader Joe's Fancy Raw
  Mixed Nuts
- Trader Joe's Go Raw
  Trek Mix
- Trader Joe's Old Fashioned
  Blister Peanuts (no salt)
- Trader Joe's Organic Fruit &
  Nut Trail Mix
- Trader Joe's Organic Raw
  Pumpkin Seeds
- Trader Joe's Organic Whole
  Filberts
- Trader Joe's Raw Nonpareil
  Almonds
- Trader Joe's Raw Pecan Pieces
- Trader Joe's Raw Pepitas
- Trader Joe's Raw Pistachio
  Nutmeats—Halves & Pieces
- Trader Joe's Raw Pistachios
  (in shell)
- Trader Joe's Raw Sunflower
  Seeds
- Trader Joe's Raw Whole
  Cashews
- Trader Joe's Roasted &
  Unsalted Peanuts
- Trader Joe's Roasted &
  Unsalted Sunflower Seeds
- Trader Joe's Roasted &
  Unsalted Whole Cashews

Packaged Fresh Berries
- California Giant Blueberries
- Sandpiper Organic Raspberries
- Sandpiper Organic
  Strawberries
    www.beachstreet.com

Packaged Fresh Berries (*continued*)

- Townsend Farms Organic Blueberries

- Townsend Farms Blackberries
www.townsendfarms.com

Packaged Greens/Vegetables

- Cut'n Clean Greens—Country Mix (collards/mustard/turnip)
    1-888-3GREENS
    www.cutncleangreens.com
- Dole Salad Mix—multiple blends
- Fresh Express-Salads— multiple blends
    1-800-242-5472
    www.freshexpress.com
- Green Giant Fresh
- Grimmway Farms Shredded Carrots
- Grimmway Farms Carrot Chips
    1-800-301-3101
    www.grimmway.com
- Mann's Sunny Shores Broccoli
- Mann's Cole Slaw (broccoli, carrots, red cabbage)
- Mann's Broccoli and Cauliflower
- Mann's Broccoli Wokly Broccoli Florettes

- Mann's Vegetable Medley (broccoli, cauliflower, baby carrots)
- Mann's Cauliettes (cauliflower florets)
- Mann's Sugar Snaps
    1-800-285-1002
    www.broccoli.com
- Organic Earthbound Farm Baby Spinach Salad— multiple blends
    www.ebfarm.com
- Packaged Greens
    Tanimura and Antle, Inc.
    Salinas, CA 93912-4070
    1-800-772-4542
    www.taproduce.com
- Ready Pac—multiple blends
    1-800-800-7822
    www.readypacproduce.com
- Red Shred (red cabbage)
- Stick Pack (carrots, celery)
    Garden Preps by: Pearson Foods Corp.
    Grand Rapids, MI 49508

Pancake and Waffle Mix

- Natural Ovens Pancake & Waffle Mix
    1-800-558-3535
    www.naturalovens.com
    They will ship products to your home.

- Arrowhead Mills Buckwheat Pancake and Waffle Mix
- Arrowhead Mills Multigrain Pancakes and Waffle Mix

Pasta

- Al Dente Spinach Fettuccine Noodles
    1-800-536-7278
    www.aldentepasta.com
- American Beauty Healthy Harvest
- American Beauty Whole Wheat Blend Pasta— Thin Spaghetti Style
- American Beauty Whole Wheat Blend Pasta— Spaghetti Style
    1-800-730-5957
- Annie's Homegrown Organic Whole Wheat Spaghetti
    www.annies.com
- Bean Cuisine—cooks in 15 minutes
- Florentine Beans with Bow Ties (includes pasta, beans, herbs, and spices)
    Reily Foods Co.
    640 Magazine St.
    New Orleans, LA 70130
    1-504-524-6131

- DeBoles Organically Produced Whole Wheat Spaghetti Style Pasta
- Organica Di Sicilia Spaghetti
- Organica Di Sicilia Fettuccine
- Organica Di Sicilia Whole Wheat Fettuccine
- Organica Di Sicilia Whole Wheat Spirali
    1-800-277-4268
- Pasta Del Verde Spaghetti
- 141 Durum Whole Wheat Pasta—Enriched Macaroni Product
    www.delverde.com
- Pasta Zesta—Garlic & Parsley Pasta
- Pasta Zesta—Tomato & Basil Pasta
    P.O. Box 7401-705
    Studio City, CA 91604
- Ryvita Flavorful Fiber Whole Grain
- Ryvita Dark Rye
- Trader Joe's Organic Linguine

Peanut Butter

- Arrowhead Mills 100% Valencia Peanut Butter—no added salt, sugar, preservatives
- Arrowhead Mills Crunchy Valencia Peanut Butter
- Arrowhead Mills Sesame Tahini
  www.arrowheadmills.com
- Maranatha Almond Butter—crunchy, no salt
  www.nspiredfoods.com/maranatha.html

- All Natural Laura Scudder's Old Fashioned Peanut Butter—smooth or chunky, unsalted Trader Joe's (and other reputable companies) has a number of nut butters—any without added salt, sugar, or preservatives are a healthy choice.

Popcorn

- Better Than Ever Premium America Popcorn
  www.gwproducts.com
- Newman's Own Organics—Organic Microwave Popcorn
- Newman's Own Organics—Pop's Corn (with natural butter flavor)

- Newman's Own Organics—Light Butter or Butter Flavor
  www.newmansownorganics.com

The above popcorn choices have no partially hydrogenated shortening and no trans-fatty acids.

Salad Dressings

- Annie's Naturals Balsamic Vinaigrette
  www.anniesnaturals.com
- Freshly Made Morgan's Dressing—French Dressing
- Freshly Made Morgan's Dressing—Balsamic Vinaigrette
  www.dressing.com

- Kirkland Balsamic Vinegar of Modena (Costco)
- La Maison Fresh Garlic Caesar Dressing
  Seaforth Creamery, Inc.
  Seaforth, Ontario, Canada
  N0K 1W0

- Oak Hill Farms Vidalia
  Onion Vinaigrette
  www.oakhillfarms.com
- Newman's Own
  Recommended Varieties:
  Balsamic Vinaigrette
  Oil and Vinegar
  Caesar

  Ranch
  Family Recipe Italian
  Lite Italian Dressing
  Lite Balsamic Vinaigrette
  Parmesano Italiano
- My favorite salad dressing is
  extra virgin olive oil (any
  brand) plus balsamic vinegar.

Salsa—Bottled

- La Victoria Red Taco Sauce
- Santa Barbara Olive Co.—
  Chunky Olive Pesto Sauce
  1-800-624-4896
  www.sbolive.com
- Tostitos All Natural Restaurant
  Style Salsa
- Tostitos All Natural Salsa

Salsa—Fresh, Refrigerated

- Santa Barbara Mango Salsa
  with Peach—Medium
  www.sbsalsa.com
- Trader Joe's Guacamango
  Salsa

Sardines

- Beach Cliff Sardines in
  Soybean Oil
  Stinson Seafood Co.
  Prospect Harbor, ME 04669
- Bela Olhau Portugal Lightly
  Smoked Sardines in Olive Oil
  Distributed by: Blue Galleon
  Newton, MA 02458
- Crown Prince One Layer
  Brisling Sardines in oil/no salt
  added, packed in soybean oil
  (Product of Scotland)
- Crown Prince Skinless &
  Boneless Sardines in Olive Oil
  (Product of Morocco)
  Imported by: Crown
  Prince, Inc.
  City of Industry, CA 91748
- King Oscar Extra Small
  Sardines in Purest Virgin Olive
  Oil
  www.kingoscar.no
- King Oscar Finest Norwegian
  Sardines in Olive Oil

Sardines (*continued*)

- Yankee Clipper Lightly Smoked Sardines in Lemon Sauce
- Yankee Clipper Lightly Smoked Sardines in Soybean Oil
- Yankee Clipper Lightly Smoked Sardines in Tomato Sauce
  Distributed by:
  Wessanen, USA
  St. Augustine, FL 32085-0410

Soups

- Health Valley 99% Fat Free Chicken Noodle Soup
- Health Valley Fat Free Corn and Vegetable Soup
- Health Valley Garden Vegetable—Fat Free Soup
- Health Valley Low Fat Chicken Broth
- Health Valley No Salt Added Beef Flavored Broth
- Health Valley Organic Soup Tomato—no salt added
- Health Valley Split Pea Soup—no salt added
- Health Valley Vegetable Soup 1-800-434-4246
- Imagine Natural Creamy Broccoli Soup
- Imagine Natural Creamy Butternut Squash Soup
- Imagine Natural Creamy Portabello Mushroom Soup
- Imagine Natural Creamy Potato Leek Soup
  www.imaginefoods.com
- Trader Joe's Low Fat Reduced Sodium Split Pea Soup

Soy Dinner Entrées

Simply Add Veggies, Cacciatore Soy Entrée Kit
www.simplyaddveggies.com

Soymilk

- Harmony Farms Light Soy Soymilk—any "light" flavor
  Harmony Farms
  P.O. Box 410
  St. Augustine, FL 32085-0410
- Kirkland by Silk Vanilla Soymilk (Costco)
- Organic Silk Soymilk
- Organic Silk Vanilla Soymilk

- Original Edensoy Extra Organic Soymilk Fortified with Beta Carotene, vitamins B_{12}, E, D, and Calcium
- Original Edensoy Organic Soymilk
- Vanilla Edensoy Extra Organic Soymilk Fortified with Beta Carotene, itamins B_{12}, E, D, and Calcium
- Pacific Soy Organic Ultra Vanilla Soymilk
- Westsoy Organic Unsweetened Soymilk
- Westsoy Plus Soymilk Vanilla

Soy Smoothies
- Hansen's Natural Soy Smoothies—multiple flavors
  www.hansens.com

Sun-Dried Tomatoes
- Premium Valley Sun California Sun Dried Tomatoes Julienne
  www.valleysun.com

Tea
- Bigelow "Constant Comment" Green Tea flavored with orange and spice
- Bigelow Green Tea
- Celestial Seasonings Decaffeinated Green Tea
- Celestial Seasonings Wellness Tea
- Celestial Seasonings Sunburst C
- Celestial Seasonings Green Tea—Organic
- Golden Green Tea
  www.traditionalmedicinals.com
- Lipton Green Tea
- Lipton Unsweetened Iced Tea
- Salada 100% Green Tea
- Twinings of London Lady Grey Green Tea
- Twinings of London Earl Grey Tea
- Twinings of London Original Earl Grey Tea

Tea (*continued*)
- Twinings of London Jasmine Green Tea

Tortillas
- Ezekiel 4:9 New Mexico Style Sprouted Grain Tortillas—certified organic www.foodforlife.com
- Henry's Marketplace 100% Stone Ground Gourmet Corn Tortillas (Ingredients: 100% stone ground corn, water, and lime)

    This is a local San Diego market, but I listed the tortilla because it is made of healthy ingredients, plus there is 1 gram of fiber per tortilla. If there is 0.0 gram of fiber in a tortilla, look for a better choice.

- La Tortilla Factory Whole Wheat Low-Fat, Low-Carb Tortillas
    1-707-586-4000 or
    1-800-446-1516
    www.latortillafactory.com
- Tumaro's Gourmet Tortillas Honey Wheat—uses organic flour
    www.tumaros.com

Veggie Hot Dogs
- The Good Dog
    1-800-667-9837
    www.yvesveggie.com
- Lightlife Smart Dogs
    1-800-769-3279
    www.lightlife.com

Whole Grain Granola
- Back to Nature Granola
    www.organicmilling.com
- Great Granola
    1-800-558-3535
    www.naturalovens.com
    They will ship products to your home.

- Health Valley Low Fat Granola—Tropical Fruit
- Health Valley Low Fat Granola—Date Almond Flavor
- Health Valley Low Fat Granola—Raisin Cinnamon
- Kashi Go Lean Crunch

Whole Grains

- Fantastic Organic Whole Wheat Couscous www.fantasticfoods.com
- Lundberg Family Farms Organic Long Grain Brown Rice
- Lundberg Family Farms Organic Short Grain Brown Rice
- Lundberg Family Farms Organic Wild Rice Blend www.lundberg.com
- Texmati Long Grain American Basmati Brown Rice 1-800-232-RICE www.riceselect.com
- Trader Joe's California Brown Aromatic Rice
- Trader Joe's Rice Trilogy
- Trader Joe's Basmati Rice Medley
- Trader Joe's Red Rice
- Trader Joe's Brown Rice Medley

Yogurt

- Alta Dena Low Fat/Non Fat Yogurt—multiple flavors Distributed by: Alta Dena Certified Dairy Inc. City of Industry, CA 91744
- Cascade Fresh Low Fat/Fat Free Yogurt—multiple flavors 1-800-511-0057 www.cascadefresh.com
- Colombo Low Fat/Non Fat Yogurt—multiple flavors
- Continental Yogurt Low Fat/ Non Fat—multiple flavors
- Horizon Organic Low Fat/ Fat Free Yogurt—multiple flavors
- Kirkland Low Fat Swiss Style Yogurt—multiple flavors (Costco)
- Stonyfield Farm Organic Low Fat Yogurt—multiple flavors
- Stonyfield Farm Non Fat Yogurt—multiple flavors
- Trader Joe's French Village Low Fat/Nonfat Yogurt— multiple flavors

Yogurt Smoothies

Stonyfield Farm Organic Smoothie—multiple flavors (my favorite is Strawberry)

Always try the lowest sodium per serving that your taste buds can tolerate; it is better to add a little salt than to start off with a high-salt canned item.

Always check the label for partially hydrogenated oils. If they are listed, don't buy the item; there is no safe amount of so-called trans fat.

Go to the market with the mindset of a hunter-gatherer. First stop is the fresh fruit–vegetable section, and then on to various whole grain products, legumes, nuts and seeds, low-fat/nonfat dairy, and so forth.

On some food labels both sodium and potassium are listed; a potassium to sodium ratio of 4:1 or more is ideal.

For information on organic products, check California Certified Organic Farmers at www.ccof.org.

Bibliography

....

How Your Diet Is Killing You

Adlercreutz, H. Western diet and Western diseases: some hormonal and biochemical mechanisms and associations. *Scand J Clin Lab Invest* 1990;201(suppl):3–23.

Albanes, D., et al. Antioxidants and cancer: evidence from human observational studies and intervention trials. In: *Antioxidant Status, Diet, Nutrition and Health*. Papas, A.M., ed. CRC Press;1999:497–544.

Albertson, A.M., et al. Consumption of grain and whole-grain foods by an American population during the years of 1990–1992. *J Am Diet Assoc* 1995;95:703–4.

Ascherio, A., et al. Intake of potassium, magnesium, calcium, and fiber and risk of stroke among US men. *Circulation* 1998;98(12):1198–204.

Bantle, J.P., et al. Effects of dietary fructose on plasma lipids in healthy subjects. *Am J Clin Nutr* 2000;72:1128–34.

Bazzano, L.A., et al. Fruit and vegetable intake and risk of cardiovascular disease in US adults: the first National Health and Nutrition Examination Survey Epidemiologic Follow-up Study. *Am J Clin Nutr* 2002;76(1):93–9.

Brand, J., et al. Food processing and the glycemic index. *Am J Clin Nutr* 1985;42:1192–6.

Centers for Disease Control and Prevention. Chronic diseases and their risk factors: the nation's leading causes of death. Atlanta: Centers for Disease Control and Prevention, 1999.

Coulston, A.M. The role of dietary fats in plant-based diets. *Am J Clin Nutr* 1999;70(suppl):512S–5S.

Davis, C.D. Diet and carcinogenesis. In: *Vegetables, Fruits, and Herbs in Health Promotion.* Watson, R.R., ed. CRC Press; 2001:273–92.

DeBoer, S.W., et al. Dietary intake of fruits, vegetables, and fat in Olmsted County, Minn. *Mayo Clinical Proceedings* 2003;78: 161–6.

Dock, W. The reluctance of physicians to admit that chronic disease may be due to faulty diet. Reprinted from 1953. *Am J Clin Nutr* 2003;77(6):1345–7.

Fontaine, K.R., et al. Years of life lost due to obesity. *JAMA* 2003;289(2):187–93.

Food and Nutrition Board, Institute of Medicine. Dietary Intake Data from the Third National Health and Nutrition Examination Survey (NHANESIII), 1988–1994. In: *Dietary Reference Intakes, appendix C.* Washington, D.C.: National Academy Press;2001:594–643.

Fraser, G.E., et al. Effect of risk factor values on lifetime risk of and age at first coronary event. The Adventist Health Study. *Am J Epidemiol* 1995;142:746–58.

Fraser, G.E., et al. Risk factors for all-cause and coronary heart disease mortality in the oldest-old. The Adventist Health Study. *Arch Intern Med* 1997;157:2249–58.

Fraser, G.E. Associations between diet and cancer, ischemic heart disease, and all-cause mortality in non-Hispanic white California Seventh-Day Adventists. *Am J Clin Nutr* 1999;70 (suppl):532S–8S.

Friedenreich, C.M. Physical activity and cancer: lessons learned from nutritional epidemiology. *Nutr Rev* 2001;59(11): 349–57.

Fung, T.T., et al. Association between dietary patterns and plasma biomarkers of obesity and cardiovascular disease risk. Am J Clin Nutr 2001;73:61–7.

Harnack, L.J., et al. Temporal trends in energy intake in the United States: an ecologic perspective. *Am J Clin Nutr* 2000; 71:1478–84.

Kanazawa, M., et al. Effects of a high-sucrose diet on body weight, plasma triglycerides, and stress tolerance. *Nutr Rev* 2003;61(5, Part II):S27–S33.

Kimura, S. Glycemic carbohydrate and health: background and synopsis of the symposium. *Nutr Rev* 2003;61(5, Part II): S1–S4.

O'Dea, K. Clinical implications of the "thrifty genotype" hypothesis: where do we stand now? *Nutr Metab Cardiovasc Dis* 1997; 7:281–4.

Ramakrishnan, U. Prevalence of micronutrient malnutrition worldwide. *Nutr Rev* 2002;60(5, Part II):S46–S52.

Rolls, B.J., et al. Portion size of food affects energy intake in normal-weight and overweight men and women. *Am J Clin Nutr* 2002;76(6):1207–13.

Rutledge, J.C. Links between food and vascular disease. *Am J Clin Nutr* 2002;75(1):4.

Micronutrients: The Keys to Super Health

Cao, G., et al. Antioxidant capacity of tea and common vegetables. *J Agric Food Chem* 1996;4:3426–31.

Hennekens, C.H. Antioxidant vitamins and cardiovascular disease. In: *Antioxidant Status, Diet, Nutrition and Health*. Papas, A.M., ed. CRC Press;1999:463–78.

Jacob, R.A., et al. Oxidative damage and defense. *Am J Clin Nutr* 1996;63(suppl):985S–90S.

O'Neill, K.L., et al. Fruits and vegetables and the prevention of oxidative DNA damage. In: *Vegetables, Fruits, and Herbs in Health Promotion*. Watson, R.R., ed. CRC Press;2001:135–46.

Phytochemicals—A New Paradigm. Bidlack, W.R., Omaye, S.T., Meskin, M.S., Jahner, D., eds. Lancaster, PA: Technomic Publishing Co. Inc., 1998.

Phytochemicals as Bioactive Agents. Bidlack, W.R., Omaye, S.T., Meskin, M.S., Jahner, D., eds. Boca Raton, FL: CRC Press LLC, 2000.

Rimm, E.B., et al. Vegetable, fruit, and cereal fiber intake and risk of coronary heart disease among men. *JAMA* 1996;275: 447–51.

Shahidi, F., et al. *Phenolics in Food and Nutraceuticals.* Boca Raton, FL: CRC Press LLC, 2004.

Simopoulos, A.P. The Mediterranean diets: what is so special about the diet of Greece? The scientific evidence. *J Nutr* 2001; 131:3065S–73S.

Trichopoulou, A., et al. Mediterranean Diet: are antioxidants central to its benefits? In: *Antioxidant Status, Diet, Nutrition and Health.* Papas, A.M., ed. CRC Press;1999:107–18.

Waladkhani, A.R., et al. Effect of dietary phytochemicals on cancer development. In: *Vegetables, Fruits, and Herbs in Health Promotion.* Watson, R.R., ed. CRC Press;2001:3–18.

Wattenberg, L.W. An overview of chemoprevention: current status and future prospects. *Proc Soc Exp Biol Med* 1997;216:133–41.

Wise, J.A. Health benefits of fruits and vegetables: the protective role of phytonutrients. In: *Vegetables, Fruits, and Herbs in Health Promotion.* Watson, R.R., ed. CRC Press;2001:147–76.

The Four Principles of SuperFoods Rx

Anatomy of an Illness as Perceived by the Patient—Reflections on Healing and Regeneration. Cousins, N, ed. New York: Norton, 1979.

Anderson, J.W. Diet first, then medication for hypercholesterolemia. *JAMA* 2003;290(4):531–8.

Appel, L.J. The role of diet in the prevention and treatment of hypertension. *Curr Atheroscler Rep* 2000;2:521–28.

Cao, G., et al. Increases in human plasma antioxidant capacity after consumption of controlled diets high in fruit and vegetables. *Am J Clin Nutr* 1998;68:1081–7.

Carson, R. *Silent Spring.* Boston, MA: Houghton Mifflin, 1994.

Cordian, L. The nutritional characteristics of a contemporary diet based upon Paleolithic food groups. *JANA* 2002;5(3):15–24.

de Lorgeril, M., et al. Mediterranean dietary pattern in a randomized trial. Prolonged survival and possible reduced cancer rate. *Arch Intern Med* 1998;158:1181–7.

de Lorgeril, M., et al. Modified Cretan Mediterranean diet in the prevention of coronary heart disease and cancer. In: *Mediterranean Diets*. Simopoulos, A.P., Visioli, F., eds. Karger Basel, Switzerland 2000;87:1–23.

Eastell, R., et al. Strategies for skeletal health in the elderly. *Proc Nutr Soc* 2002;61(2):173–80.

Fairfield, K.M., et al. Vitamins for chronic disease prevention in adults: scientific review. *JAMA* 2002;287(23):3116–26.

Fan, W.Y., et al. Reduced oxidative DNA damage by vegetable juice intake: a controlled trial. *J Physiol Anthropol Appl Human Sci* 2000;19:287–9.

Fleet, J.C. DASH without the dash (of salt) can lower blood pressure. *Nutr Rev* 2001;59(9):291–7.

Gronbaek, M., et al. Type of alcohol consumed and mortality from all causes, coronary heart disease, and cancer. *Ann Intern Med* 2000;133:411–9.

Gussow, J.D. *This Organic Life: Confessions of a Suburban Homesteader.* White River Junction, VT: Chelsea Green Publishing Company, 2001.

Jenkins, D.J.A., et al. Effects of a dietary portfolio of cholesterol-lowering foods vs lovastatin on serum lipids and c-reactive protein. *JAMA* 2003;290(4):502–10.

Kant, A.K., et al. Dietary diversity and subsequent mortality in the First National Health and Nutrition Survey Epidemiologic Follow-up Study. *Am J Clin Nutr* 1993;57:434–40.

Marlett, J.A., et al. Position of the American Dietetic Association: health implications of dietary fiber. *J Am Diet Assoc* 2002; 102(7):993–1000.

Milton, K. Nutritional characteristics of wild primate foods: do the natural diets of our closest living relatives have lessons for us? *Nutr* 1999;15:488–98.

Mozaffarian, D., et al. Cereal, fruit, and vegetable fiber intake and the risk of cardiovascular disease in elderly individuals. *JAMA* 2003;289(13):1659–66.

Nick, G.L. Detoxification properties of low-dose phytochemical complexes found within select vegetables. *JANA* 2002;5(4): 34–44.

Pool-Zobel, B.L., et al. Consumption of vegetables reduces genetic damage in humans: first results of a human intervention trial with carotenoids-rich foods. *Carcinogenesis* 1997; 18:1847–50.

Pratt, et al. Nutrition and Skin Cancer Risk Prevention. In: *Functional Foods and Neutraceuticals in Cancer Prevention*, Ronald R. Watson, editor, Iowa State Press, 2003:105–20.

Simopoulos, A.P. Essential fatty acids in health and chronic disease. *Am J Clin Nutr* 1999;70(suppl):560S–9S.

Storper, B. Moving toward healthful sustainable diets. *Nutr Today* 2003;38(2):57–9.

Sun, J., et al. Antioxidant and antiproliferative activities of common fruits. *J Agric Food Chem* 2002;50(25):7449–54.

Suter, P.M. Alcohol and mortality: if you drink, do not forget fruits and vegetables. *Nutr Rev*; 59(9):293–7.

Thompson, H.J., et al. Effect of increased vegetable and fruit consumption on markers of oxidative cellular damage. *Carcinogenesis* 1999;20:2261–6.

USDA Nutrient Database for Standard Reference. http://www.nal.usda.gov/fnic/cgi-bin/nut_search.pl

White, I.R. The level of alcohol consumption at which all-cause mortality is least. *J Clin Epidemiol* 1999;52:967–75.

Willcox, B.J. *The Okinawa Program.* Willcox, B.J., Willcox, D.C., Suzuki, M., eds. New York: Three Rivers Press, 2001.

Willett, W.C., et al. Mediterranean diet pyramid: a cultural model for healthy eating. *Am J Clin Nutr* 1995;61(suppl):1402S–6S.

Willett, W.C., et al. Relation of meat, fat, and fiber intake to the risk of colon cancer in a prospective study among women. *N Engl J Med* 1990;323:1664–72.

Willett, W.C. *Eat, Drink and Be Healthy—The Harvard Medical School Guide to Healthy Eating.* New York: Simon & Schuster Source, 2001.

Willett, W.C. Micronutrients and cancer risk. *Am J Clin Nutr* 1994;59:162S–5S.

SuperFoods Rx in Your Kitchen

American Institute for Cancer Research. As restaurant portions grow, vast majority of Americans still belong to "clean plate club," new survey finds. January 15, 2001. American Institute for Cancer Research home page: www.aicr.org. Internet: http://www.aicr.org/r011501.htm (accessed 8 November 2001).

American Institute for Cancer Research. New survey shows Americans ignore importance of portion size in managing weight. March 24, 2000. American Institute for Cancer Research home page: www.aicr.org. Internet: http://www.aicr.org/r032400.htm (accessed 8 November 2001).

Duyff, R.L. *American Dietetic Association Complete Food and Nutrition Guide*, 2nd edition. Hoboken, NJ: John Wiley & Sons, Inc., 2002.

Freedman, D.S., et al. Trends and correlates of class 3 obesity in the United States from 1990 through 2000. *JAMA* 2002; 288(14):1758–61.

Hu, F.B., et al. Optimal diets for prevention of coronary heart disease. *JAMA* 2002;299(20):2569–78.

Kant, A.K., et al. A prospective study of diet quality and mortality in women. *JAMA* 2000;283:2109–15.

Krebs-Smith, S.M., et al. The effects of variety in food choices on dietary quality. *J Am Diet Assoc* 1987;87(7):896–903.

Newby, P.K., et al. Dietary patterns and changes in body mass index and waist circumference in adults. *Am J Clin Nutr* 2003; 77(6):1417–25.

Ogden, C.L., et al. Prevalence and trends in overweight among US children and adolescents, 1999–2000. *JAMA* 2002;288(14): 1728–32.

Beans

Adlercreutz, H.A., et al. Effect of dietary components, including lignins and phytoestrogens on enterohepatic circulation and live metabolism of estrogens and on sex hormone binding globulin. *J Steroid Biochem* 1987;27:1135–44.

The American Cancer Society, Dietary Guidelines Advisory Committee. Guidelines on diet, nutrition and cancer prevention:

reducing the risk of cancer with healthy food choices and physical activity. 1996.

Anderson, J.W., et al. Cardiovascular and renal benefits of dry bean and soybean intake. *Am J Clin Nutr* 1999;70(3 suppl): 464S–74S.

Anderson, J.W., et al. Hypocholesterolemic effects of oat and bean products. *Am J Clin Nutr* 1988;48:749–53.

Barampama, Z., et al. Oligosaccharides, antinutritional factors, and protein digestibility of dry beans as affected by processing. *J Food Sci* 1994;59:833–8.

Bazzano, L.A., et al. Legume consumption and risk of coronary heart disease in US men and women: NHANES I Epidemiologic Follow-up Study. *Arch Int Med* 2001;161(21):2573–8.

Brown, L., et al. Cholesterol-lowering effects of dietary fiber; a meta-analysis. *Am J Clin Nutr* 1999;69:30–42.

Correa, P. Epidemiologic correlations between diet and cancer frequency. *Cancer Res* 1981;41:3685–9.

Deshpande, S.S. Food legumes in human nutrition: a personal perspective. *CRC Crit Rev Food Sci Nutr* 1992;32:333–63.

Geil, P.B., et al. Nutrition and health implications of dry beans: a review. *J Am Coll Nutr* 1994;13:549–58.

Graf, E., et al. Suppression of colonic cancer by dietary phytic acid. *Nutr Cancer* 1993;19:11–9.

Jenkins, D.J.A., et al. Exceptionally low blood glucose response to dried beans: comparison with other carbohydrate foods. *Br Med J* 1980;281:578–80.

Kushi, L.H., et al. Cereals, legumes, and chronic disease risk reduction: evidence from epidemiologic studies. *Am J Clin Nutr* 1999;70(suppl):451S–8S.

McIntosh, M. A diet containing food rich in soluble and insoluble fiber improves glycemic control and reduces hyperlipidemia among patients with type 2 diabetes mellitus. *Nutr Rev* 2001;59(2):52–5.

Mazur, W., et al. Isoflavonoids and lignans in legumes: nutritional and health aspects in humans. *J Nutr Biochem* 1998;9:193–200.

Menotti, A., et al. Food intake patterns and 25-year mortality from coronary heart disease: cross-cultural correlations in the Seven

Countries Study. The Seven Countries Study Research Group. *Eur J Epidemiol* 1999;15(6):507–15.

Miller, J.W. Does lowering plasma homocysteine reduce vascular disease risk? *Nutr Rev* 2001;59(7):242–4.

Morrow, B. The rebirth of legumes. *Food Technol* 1991;45(4): 96–101.

Schafer, G., et al. Comparison of the effects of dried peas with those of potatoes in mixed meals on postprandial glucose and insulin concentrations in patients with type 2 diabetes. *Am J Clin Nutr* 2003;78(1):99–103.

Shutler, S.M., et al. The effect of daily baked bean (Phaseolus vulgaris) consumption on the plasma lipid levels of young, normo-cholesterolemic men. *Br J Nutr* 1989;61:257–63.

Slattery, M.L., et al. Plant foods and colon cancer: an assessment of specific foods and their related nutrients (United States). *Cancer Causes Control* 1997;8:575–90.

van Horn, L. Fiber, lipids, and coronary heart disease: a statement for healthcare professionals from the Nutrition Committee, American Heart Association. *Circulation* 1997;95:2701–4.

Vinson, J.A., et al. Phenol antioxidant quantity and quality in foods: vegetables. *J Agric Food Chem* 1998;46(9):3630–34.

Blueberries

Amakura, Y., et al. Influence of jam processing on the radical scavenging activity and phenolic content in berries. *J Agric Food Chem* 2000;48(12):6292–7.

Bravo, L. Polyphenols: chemistry, dietary sources, metabolism, and nutritional significance. *Nutr Rev* 1998;56(11):317–33.

Cao, G., et al. Anthocyanins are detected in human plasma after oral administration of an elderberry extract. *Clin Chem* 1999;45:574–6.

Cao, G., et al. Serum antioxidant capacity is increased by consumption of strawberries, spinach, red wine or vitamin C in elderly women. *J Nutr* 1998;128(12):2383–90.

Commenges, D., et al. Intake of flavonoids and risk of dementia. *Eur J Epidemiol* 2000;16:357–63.

Das, D.K., et al. Cardioprotection of red wine: role of polyphenolic antioxidants. *Drugs Exp Clin Res* 1999;25(2–3):115–20.

Erlund, I., et al. Consumption of black currants, lingonberries and bilberries increases serum quercetin concentrations. *Eur J Clin Nutr* 2003;57(1):37–42.

Ferrandiz, M.L., et al. Anti-inflammatory activity and inhibition of arachidonic acid metabolism by flavonoids. *Agents Actions* 1991;32:283–8.

Fuhrman, B., et al. Consumption of red wine with meals reduces the susceptibility of human plasma and low-density lipoprotein to lipid peroxidation. *Am J Clin Nutr* 1995;61:549–54.

Gheldof, N., et al. Buckwheat honey increases serum antioxidant capacity in humans. *J Agric Food Chem* 2003;51(5):1500–5.

Gil, M.I., et al. Antioxidant activity of pomegranate juice and its relationship with phenolic composition and processing. *J Agric Food Chem* 2000;48(10):4581–9.

Girard, B., et al. Functional grape and citrus products. In: *Functional Foods Biochemical & Processing Aspects*. Mazza, G., ed. Lancaster, PA: Technomic Publishing Company, Inc., 1998:139–92.

Hertog, M.G.L., et al. Antioxidant flavonols and coronary heart disease risk. *Lancet* 1997;349:699.

Hertog, M.G.L., et al. Antioxidant flavonols and ischemic heart disease in a Welsh population of men: the Caerphilly study. *Am J Clin Nutr* 1997;65:1489–94.

Hertog, M.G.L., et al. Dietary antioxidant flavonoids and risk of coronary heart disease: the Zutphen elderly study. *Lancet* 1993;342:1007–11.

Hertog, M.G.L., et al. Dietary flavonoids and cancer risk in the Zutphen Elderly Study. *Nutr Cancer* 1994;22:175–84.

Joseph, J.A., et al. Long-term dietary strawberry, spinach, or vitamin E supplementation retards the onset of age-related neuronal signal-transduction and cognitive behavioral deficits. *J Neurosci* 1998;18(19):8047–55.

Joseph, J.A., et al. Oxidative stress protection and vulnerability in aging: putative nutritional implications for intervention. *Mech Ageing Dev* 2000;31;116(2–3):141–53.

Joseph, J.A., et al. Reversals of age-related declines in neuronal signal transduction, cognitive, and motor behavioral deficits with blueberry, spinach, or strawberry dietary supplementation. *J Neurosci* 1999;19(18):8114–21.

Kay, C.D., et al. The effect of wild blueberry (*Vaccinium angustifolium*) consumption on postprandial serum antioxidant status in human subjects. *Br J Nutr* 2002;88(4):389–98.

Keevil, J.G., et al. Grape juice, but not orange juice or grapefruit juice, inhibits human platelet aggregation. *J Nutr* 2000;130(1): 53–6.

Knekt, P., et al. Dietary flavonoids and the risk of lung cancer and other malignant neoplasms. *Am J Epidemiol* 1997;146:223–30.

Kopp, P. Resveratrol, a phytoestrogen found in red wine: a possible explanation for the conundrum of the French paradox? *Eur J Endocrinol* 1999;138(6):619–20.

Lansky, E., et al. Pharmacological and therapeutical properties of pomegranate. In: *Proceedings 1st International Symposium on Pomegranate;* Megarejo, P. Martinez, J. J. Martinez J., eds., CIHEAM, Orihuela, Spain, 1998;Pr–07.

Maas, J.L., et al. Ellagic acid, an anticarcinogen in fruits, especially in strawberries: a review. *Hortic Sci* 1991;26:10–14.

Mazza, G., et al. Absorption of anthocyanins from blueberries and serum antioxidant status in human subjects. *J Agric Food Chem* 2002;50(26):7731–7.

O'Byrne, D.J., et al. Comparison of the antioxidant effects of Concord grape juice flavonoids and α-tocopherol on markers of oxidative stress in healthy adults. *Am J Clin Nutr* 2002; 76(6):1367–74.

Oregon Berries Web Page: http://www.oregon-berries.com

Paper, D.H. Natural products as angiogenesis inhibitors. *Planta Med* 1998;64:686–95.

Saija, A., et al. Flavonoids as antioxidant agents: importance of their interaction with biomembranes. *Free Radic Biol Med* 1995;19(4):481–6.

Stacewicz-Sapuntzakis, M., et al. Chemical composition and potential health effects of prunes: a functional food? *Crit Rev Food Sci Nutr* 2001;41(4):251–86.

USDA Database for the Flavonoid Content of Selected Foods. Accessed March 2003. www.nalusda.gov/fnic/foodcomp

Vinson, J.A. Total polyphenol content of selected juices and jams. Personal communication, 2003.

Vinson, J.A., et al. Phenol antioxidant quantity and quality in foods: fruits. *J Agric Food Chem* 2001;49(11):5315–21.

Broccoli

Cohen, J.H., et al. Fruit and vegetable intakes and prostate cancer risk. *J Natl Cancer Inst* 2000;92(1):61–8.

Conaway, C.C., et al. Disposition of glucosinolates and sulforaphane in humans after ingestion of steamed and fresh broccoli. *Nutr Cancer* 2000;38(2):168–78.

Ernster, L., et al. Biochemical, physiological and medical aspects of ubiquinone function. *Biochem Biophys Acta* 1995;1271: 195–204.

Fahey, J.W., et al. Antioxidant functions of sulforaphane: a potent inducer of phase 2 detoxication enzymes. *Food Chem Toxicol* 1999;37:973–9.

Jeffery, E.H., et al. Cruciferous Vegetables and Cancer Prevention. In: *Handbook of Nutraceuticals and Functional Foods.* Wildman, R.E.C., ed. Boca Raton, FL: 2001;169–92. CRCS Press LLC.

Michnovicz, J.J., et al. Altered estrogen metabolism and excretion in humans following consumption of I3C. *Nutrition and Cancer* 1991;16:59–66.

Murray, S., et al. Effect of cruciferous vegetable consumption on heterocyclic aromatic amine metabolism in man. *Carcinogenesis* 2001;22(9):1413–20.

Nestle, M. Broccoli sprouts in cancer prevention. *Nutr Rev* 1998;56(4 Pt 1):127–30.

Osborne, M.P. Chemoprevention of breast cancer. *Surg Clin North Am* 1999;79(5):1207–21.

Telang, N.T., et al. Inhibition of proliferation and modulation of estradiol metabolism: novel mechanisms for breast cancer prevention by the phytochemical indole-3-carbinol. *Proc Soc Exp Biol Med* 1997;216:246–52.

Van Poppel, G., et al. Brassica vegetables and cancer prevention. Epidemiology and mechanisms. *Adv Exp Med Biol* 1999; 472:159–68.

Verhagen, J., et al. Reduction of oxidative DNA-damage in humans by Brussels sprouts. *Carcinogenesis* 1995;16(4): 969–70.

Verhoeven, D.E., et al. A review of mechanisms underlying anti-carcinogenicity by brassica vegetables. *Chem Biol Interact* 1997;103(2):79–129.

Verhoeven, D.T.H., et al. Epidemiological studies on brassica vegetables and cancer risk. *Cancer Epidemiol Bio Prev* 1996; 5(9):733–48.

Wattenberg, L.W. Inhibition of carcinogenesis by minor anutrient constituents of the diet. *Proc Nutr Soc* 1990;49:173–83.

Oats

Albertson, A., et al. Consumption of grain and whole-grain foods by an American population during the years 1990–1992. *J Am Diet Assoc* 1995;95:703–4.

Boushey, C.J., et al. A quantitative assessment of plasma homo-cysteine as a risk factor for vascular disease. Probable benefit of increasing folic acid intakes. *JAMA* 1995;274:1049–57.

Cleveland, L.E., et al. Dietary intake of whole grains. *J Am Coll Nutr* 2000;19:331S–8S.

Cunnane, S.C. Metabolism and function of a-linolenic acid in humans. In: *Flaxseed in Human Nutrition.* Cunnane, S.C., and Thompson, L.U., eds. Champaign, IL: AOCS Press, 1995:99–127.

Cunnane, S.C., et al. Nutritional attributes of traditional flaxseed in healthy young adults. *Am J Clin Nutr* 1995;61(1):62–8.

de Lorgeril, M., et al. Mediterranean alpha-linolenic acid-rich diet in secondary prevention of coronary heart disease. *Lancet* 1994;343(8911):1454–9.

Fung, T.T., et al. Whole-grain intake and the risk of type 2 diabetes: a prospective study in men. *Am J Clin Nutr* 2002; 76(3):535–40.

Jacobs, D.R., et al. Is whole grain intake associated with reduced total and cause-specific death rates in older women? The Iowa Women's Health Study. *Am J Public Health* 1999;89:322–9.

Jacobs, D.R., et al. Whole grain intake and cancer: an expanded review and meta-analysis. *Nutr Cancer* 1998;30(2):85–96.

Jacobs, D.R., et al. Whole-grain intake may reduce the risk of ischemic heart disease death in postmenopausal women: the Iowa Women's Health Study. *Am J Clin Nutr* 1998:68(2): 248–57.

Johnston, L., et al. Cholesterol-lowering benefits of a whole grain oat ready-to-eat cereal. *Nutr Clin Care* 1998;1:6–12.

Katz, D.L., et al. Acute effects of oats and vitamin E on endothelial responses to ingested fat. *Am J Prev Med* 2001;20(2): 1124–9.

Kilkkinen, A., et al. Research Communication: intake of lignins is associated with serum enterolactone concentration in Finnish men and women. *J Nutr* 2003;133(6):1830–3.

Lampi, A.M., et al. Tocopherols and tocotrienols from oil and cereal grains. In: *Functional Foods: Biochemical and Processing Aspects*, vol. 2. Shi, J., Mazza, G., Le Maguer. M., eds. Boca Raton, FL: CRC Press, LLC, 2002: 1–38.

Levine, A.S., et al. Dietary fiber: does it affect food intake and body weight? In: *Appetite and Body Weight Regulation: Sugar, fat and macronutrient substitutes*. Fernstrom, J.D., Miller, G.D., eds. Boca Raton, FL: CRC Press, Inc., 1994:191–200.

Liu, S., et al. Is intake of breakfast cereals related to total and cause-specific mortality in men? *Am J Clin Nutr* 2003; 77(3):594–9.

Liu, S., et al. Relation between a diet with a high glycemic load and plasma concentrations of high-sensitivity C-reactive protein in middle-aged women. *Am J Clin Nutr* 2002;75(3):492–8.

Liu, S., et al. Whole grain consumption and risk of ischemic stroke in women; a prospective study. *JAMA* 2000;284: 1534–40.

Liu, S., et al. Whole-grain consumption and risk of coronary heart disease: results from the Nurses' Health Study. *Am J Clin Nutr* 1999;70(3):412–9.

McKeown, N.M., et al. Whole grain intake and risk of ischemic stroke in women. *Nutr Rev* 2001;59(5):149–152.

Miller, H.E., et al. Antioxidant content of whole grain breakfast cereals, fruits and vegetables. *J Am Coll Nutr* 2000;19(suppl): 312S–9S.

Montonen, J., et al. Whole-grain and fiber intake and the incidence of type 2 diabetes. *Am J Clin Nutr* 2003;77(3):622–9.

Oomah, B.D., et al. Flaxseed products for disease prevention. In: *Functional Foods Biochemical & Processing Aspects*. Mazza, G, ed. Lancaster, PA: Technomic Publishing Company, Inc., 1998:91–138.

Pedersen, B., et al. Nutritive value of cereal products with emphasis on the effect of milling. *World Rev Nutr Diet* 1989;60:1–91.

Saltzman, E., et al. An oat-containing hypocaloric diet reduces systolic blood pressure and improves lipid profile beyond effects of weight loss in men and women. *J Nutr* 2001; 131:1465–70.

Slavin, J., et al. Grain processing and nutrition. *Crit Rev Food Sci Nutr* 2000;40:309–26.

Slavin, J., et al. Plausible mechanisms for the protectiveness of whole grains. *Am J Clin Nutr* 1999;70(3 suppl):459S–63S.

Slavin, J., et al. Whole grain consumption and chronic disease: protective mechanisms. *Nutr Cancer* 1997;27:14–21.

Thompson, L.U. Antioxidants and hormone-mediated health benefits of whole grains. *Crit Rev Food Sci Nutr* 1994;34(586): 473–97.

Tousoulis, D., et al. L-arginine in cardiovascular disease: dream or reality? *Vasc Med* 2002;7(3):203–11.

Trusswell, A.S. Cereal grains and coronary heart disease. *Eur J Clin Nutr* 2002;56(1):1–14.

Oranges

Amparo, C., et al. Limonene from citrus. In: *Functional Foods: Biochemical and Processing Aspects*, vol. 2. Shi, J., Mazza, G., Le Maguer, M., eds. CRC Press LLC 2002: 169–88.

Block, G., et al. Ascorbic acid status and subsequent diastolic and systolic blood pressure. *Hypertension* 2001;37:261–67.

Crowell, P.L. Prevention and therapy of cancer by dietary monoterpenes. *J Nutr* 1999;129(3):775S–8S.

Gey, K.F., et al. Increased risk of cardiovascular disease at suboptimal plasma concentrations of essential antioxidants: an epidemiological update with special attention to carotene and vitamin C. *Am J Clin Nutr* 1993;57(suppl):787S–97S.

Gil-Izquierdo, A., et al. Effect of processing techniques at industrial scale on orange juice antioxidant and beneficial health compounds. *J Agric Food Chem* 2002;50(18):5107–14.

Girard, B., et al. Functional grape and citrus products. In: *Functional Foods Biochemical & Processing Aspects.* Mazza, G., ed., Lancaster, PA: Technomic Publishing Company, Inc. 1998:139–92.

Hakim, I.A., et al. Citrus peel use is associated with reduced risk of squamous cell carcinoma of the skin. *Nutr Cancer* 2000;37(2):161–8.

Hallberg, L., et al. Effect of ascorbic acid on iron absorption from different types of meals. Studies with ascorbic acid rich roods and synthetic ascorbic acid given in different amounts in different meals. *Hum Nutr Appl Nutr* 1986;40:97–113.

Halliwell, B. Vitamin C and genomic stability. *Mutat Res* 2001;475(1–2):29–35.

Jacques, P.F., et al. Long-term vitamin C supplement use and prevalence of early age-related lens opacities. *Am J Clin Nutr* 1997;66:911–6.

Johnston, C.S., et al. People with marginal vitamin C status are at high risk of developing vitamin C deficiency. *J Am Diet Assoc* 1999;99(7):854–6.

Johnston, C.S., et al. Stability of ascorbic acid in commercially available orange juices. *J Am Diet Assoc* 2002;102:525–9.

Leonard, S.S., et al. Antioxidant properties of fruit and vegetable juices: more to the story than ascorbic acid. *Ann Clin Lab Sci* 2002;332(2):193–200.

Levine, M., et al. Criteria and recommendations for vitamin C intake. *JAMA* 1999;281(15):1415–23.

Liu, L., et al. Vitamin C preserves endothelial function in patients with coronary heart disease after a high-fat meal. *Clin Cardiology* 2002;25:219–24.

Loria, C.M., et al. Vitamin C status and mortality in US adults. *Am J Clin Nutr* 2000;72(1):139–45.

Martini, L., et al. Relative bioavailability of calcium-rich dietary sources in the elderly. *Am J Clin Nutr* 2002;76(6):1345–50.

Nagy, S. Vitamin C contents of citrus fruit and their products: a review. *J Agric-Food Chem* 1980;28(1):8–18.

Proteggcnte, A.R., et al. The antioxidant activity of regularly consumed fruit and vegetables reflects their phenolic and vitamin C composition. *Free Radic Res* 2002;36(2):217–33.

Tangpricha, V., et al. Fortification of orange juice with vitamin D: a novel approach for enhancing vitamin D nutritional health. *Am J Clin Nutr* 2003;77(6):1478–83.

Vinson, J.A., et al. In vitro and in vivo lipoprotein antioxidant effect of a citrus extract and ascorbic acid on normal and hypercholesterolemic human subjects. *J Med Food* 2001;4(4):187–92.

Wang, Q., et al. Pectin from fruits. In: *Functional Foods: Biochemical and Processing Aspects*, vol. 2. Shi, J., Mazza, G., Le Maguer M., eds., CRC Press LLC 2002: 263–310.

Pumpkin

Albanes, D., et al. Alpha-tocopherol and beta carotene supplements and lung cancer incidence in the Alpha-Tocopherol, Beta-Carotene Cancer Prevention Study: Effects of base-line characteristics and study compliance. *J Nutr Cancer Inst* 1996;88:1560–70.

The Alpha-Tocopherol, Beta Carotene Cancer Prevention Study Group. The effect of vitamin E and beta carotene on the incidence of lung cancer and other cancers in male smokers. *N Engl J Med* 1994;330:1029–35.

Ascherio, A., et al. Relation of consumption of vitamin E, vitamin C, and carotenoids to risk for stroke among men in the United States. *Ann Intern Med* 1999;130(12):963–70.

Bowen, P.E., et al. Variability of serum carotenoids in response to controlled diets containing six servings of fruits and vegetables per day. *Ann NY Acad Sci* 1993;691:241–3.

Cooper, D.A., et al. Dietary carotenoids and certain cancers, heart disease, and age-related macular degeneration: a review of recent research. *Nutr Rev* 1999;57:201–14.

D'Odorico, A., et al. High plasma levels of alpha- and beta-carotene are associated with a lower risk of atherosclerosis: results from the Bruneck study. *Atherosclerosis* 2000;153(1): 231–9.

Erdman, J.W., Jr. Variable bioavailability of carotenoids from vegetables. *Am J Clin Nutr* 1999;70:179–80.

Kohlmeier, L., et al. Epidemiologic evidence of a role of carotenoids in cardiovascular disease prevention. *Am J Clin Nutr* 1995;62(suppl):1370S–6S.

Krinsky, N.I. The antioxidant and biological properties of the carotenoids. *Ann NY Acad Sci* 1998;854:443–7.

McVean, M., et al. Oxidants and antioxidants in Ultraviolet-induced nonmelanoma skin cancer. In: *Antioxidant Status, Diet, Nutrition and Health.* Papas, A.M., ed. CRC Press 1999;401–30.

Mayne, S.T. β-Carotene, carotenoids and disease prevention in humans. *FASEB J* 1996;10:690–701.

Omenn, G.S., et al. Effects of a combination of beta carotene and vitamin A on lung cancer and cardiovascular disease. *N Engl J Med* 1996;334:1150–5.

Rock, C.L., et al. Responsiveness of serum carotenoids to a high-vegetable diet intervention designed to prevent breast cancer recurrence. *Cancer Epidemiol Biomarkers Prev* 1997;6: 617–23.

Rock, C.L. Carotenoid update. *J Am Diet Assoc* 2003;103(4): 423–5.

Stahl, W., et al. Carotenoids and carotenoids plus vitamin E protect against ultraviolet light-induced erythema in humans. *Am J Clin Nutr* 2000;71(3):795–8.

van Poppel, G., et al. Epidemiologic evidence for beta-carotene and cancer prevention. *Am J Clin Nutr* 1995;62(suppl): 1393S–1402S.

White, W.S., et al. Ultraviolet light-induced reductions in plasma carotenoids levels. *Am J Clin Nutr* 1988;47:879–83.

Yeum, K.J., et al. Carotenoid bioavailability and bioconversion. In: *Annual Review of Nutrition*, vol. 22, 2002. McCormick, D.B., Bier, D.M., Cousins, R.J., eds. 2002;483–504.

Wild Salmon

Albert, C.M., et al. Fish consumption and risk of sudden cardiac death. *JAMA* 1998;279:23–8.

Bhatnagar, D., et al. Omega-3 fatty acids: their role in the prevention and treatment of atherosclerosis related risk factors and complications. *Int J Clin Pract* 2003;57(4):305–14.

Calvo, M.S., et al. Prevalence of vitamin D insufficiency in Canada and the United States: importance to health status and efficacy of current food fortification and dietary supplement use. *Nutr Rev* 2003;61(3):107–13.

Canada Food Inspection Fact Sheet. Food safety facts on mercury and fish consumption. Available at: http://www.inspection.gc.ca/english/corpaffr/foodfacts/mercurye/shtml. Accessed June 26, 2002.

Conquer, J., et al. Human health effects of docosahexaenoic acid. In: *Functional Foods: Biochemical and Processing Aspects*, vol. 2. Shi, J., Mazza, G., Le Maguer, M., eds. CRC Press LLC 2002: 311–30.

Cunningham-Rundles, S. Is the fatty acid composition of immune cells the key to normal variations in human immune response? *Am J Clin Nutr* 2003;77(5):1096–7.

Dewailly, E., et al. Cardiovascular disease risk factors and n-3 fatty acid status in the adult population of James Bay Cree. *Am J Clin Nutr* 2002;76(1):85–92.

Dewailly, E., et al. n-3 Fatty acids and cardiovascular disease risk factors among the Inuit of Nunavik. *Am J Clin Nutr* 2001; 74(4):464–73.

Doleckk, T.A., et al. Dietary polyunsaturated fatty acids and mortality in the multiple risk factor intervention trial (MRFIT). *World Rev Nutr Diet* 1991;66:205–16.

Foulke, J.E. Mercury in fish: cause for concern? *FDA Consumer*, Sept. 1994. Accessed on FDA Website at http://www.fda.gov/fdac/reprints/mercury.html.

Freedman, D.M., et al. Sunlight and mortality from breast, ovarian, colon, prostate, and non-melanoma skin cancer: a composite death certificate based case-control study. *Occ Environ Med* 2002;59:257–62.

Freeman, M.P. Omega-3 fatty acids in psychiatry: a review. *Ann Clin Psychiat* 2000;2(3):159–65.

Harris, W.S. n-3 Long-chain polyunsaturated fatty acids reduce risk of coronary heart disease death: extending the evidence to the elderly. *Am J Clin Nutr* 2003;77(2):279–80.

Holick, M.F. Vitamin D: the underappreciated D-lightful hormone that is important for skeletal and cellular health. *Curr Opin Endocrinol Diabetes* 2002;9:87–98.

Jones, P.J.H., et al. Effect of n-3 polyunsaturated fatty acids on risk reduction of sudden death. *Nutr Rev* 2002;60(12):407–9.

Kew, S., et al. Relation between the fatty acid composition of peripheral blood mononuclear cells and measures of immune cell function in healthy, free-living subjects aged 25–72 y. *Am J Clin Nutr* 2003;77(5):1278–86.

Kris-Etherton, P.M., et al. Fish consumption, fish oil, omega-3 fatty acids and cardiovascular disease. *Circulation* 2002; 106:2747–57.

Lemaitre, R.N., et al. n-3 Polyunsaturated fatty acids, fatal ischemic heart disease, and nonfatal myocardial infarction in older adults: the Cardiovascular Health Study. *Am J Clin Nutr* 2003;77(2):319–25.

Lewis, N.M., et al. Enriched eggs as a source of n-3 polyunsaturated fatty acids for humans. *Poult Sci* 2000;79(7):971–4.

Morris, M.C., et al. Does fish oil lower blood pressure? A meta-analysis of controlled trials. *Circulation* 1993;88:523–33.

Rose, D.P., et al. Omega-3 fatty acids as cancer chemopreventive agents. *Pharmacol Ther* 1999;83(3):217–44.

Stoll, A.L. *The Omega 3 Connection*. New York: Simon & Schuster, 2001, 24.

Terry, P.D., et al. Intakes of fish and marine fatty acids and the risks of cancers of the breast and prostate and of other hormone-related cancers: a review of the epidemiologic evidence. *Am J Clin Nutr* 2003;77(3):532–43.

Tidow-Kebritchi, S., et al. Effects of diets containing fish oil and vitamin E on rheumatoid arthritis. *Nutr Rev* 2001;59(10): 335–7.

Vanschoonbeek, K., et al. Fish oil consumption and reduction of arterial disease. *J Nutr* 2003;133(3):657–60.

Woodman, R.J., et al. Effects of purified eicosapentaenoic and docosahexaenoic acids on glycemic control, blood pressure, and serum lipids in type 2 diabetic patients with treated hypertension. *Am J Clin Nutr* 2002;76(5):1007–15.

Ziboh, V.A. The significance of polyunsaturated fatty acids in cutaneous biology. *Lipids* 1996;31(suppl):S249–S53.

Soy

Adlercreutz, H.H., et al. Plasma concentrations of phytoestrogens in Japanese men. *Lancet* 1993;342:1209–10.

Anderson, J.W., et al. Soy foods and health promotion. In: *Vegetables, Fruits, and Herbs in Health Promotion*. Watson, R.R., ed. CRC Press; 2001:117–34.

Anderson, J.W. Meta-analysis of the effects of soy protein intake on serum lipids. *N Engl J Med* 1995;333(5):276–82.

Bhathena, S.J., et al. Beneficial role of dietary phytoestrogens in obesity and diabetes. *Am J Clin Nutr* 2002;76(6):1191–1201.

Chang, S.K.C. Isoflavones form soybeans and soy foods. In: *Functional Foods: Biochemical and Processing Aspects*, vol. 2. Shi, J., Mazza, G., Le Maguer, M., eds. CRC Press LLC 2002: 39–70.

Dwyer, J.T., et al. Tofu and soy drinks contain phytoestrogens. *J Am Diet Assoc* 1994;94(7):739–43.

Erdman, J.W., Jr. AHA science advisory: soy protein and cardiovascular disease: a statement for healthcare professionals from the Nutrition Committee of the AHA. *Circulation* 2000; 102(20):2555–9.

Haub, M.D., et al. Effect of protein source on resistive-training-induced changes in body composition and muscle size in older men. *Am J Clin Nutr* 2002;76(3):511–7.

Jenkins, D.J.A., et al. Effects of high- and low-isoflavone soy-foods on blood lipids, oxidized LDL, homocysteine, and blood pressure in hyperlipidemic men and women. *Am J Clin Nutr* 2002;76(1):365–72.

Kreijkamp-Kaspers, S., et al. Phyto-oestrogens and cognitive function. In: *Performance Functional Foods*. Watson, D.H., ed. CRC Press, Woodhead Publishing Limited, 2003:61–77.

Matvienko, O.A, et al. A single daily dose of soybean phytosterols in ground beef decreases serum total cholesterol and LDL cholesterol in young, mildly hypercholesterolemic men. *Am J Clin Nutr* 2002;76(1):57–64.

Messina, M., et al. Provisional recommended soy protein and isoflavones intakes for healthy adults. *Nutr Today* 2003;38(3): 100–9.

Messina, M.J. Emerging evidence on the role of soy in reducing prostate cancer risk. *Nutr Rev* 2003;61(4):117–31.

Messina, M. Legumes and soybeans: overview of their nutritional profiles and health effects. *Am J Clin Nutr* 1999;70(3 suppl): 439S–50S.

Messina, M.J., et al. Soy intake and cancer risk: a review of the in vitro and in vivo data. *Nutrition and Cancer* 1994; 21(2) 113–31.

Munro, I.C., et al. Soy isoflavones: a safety review. *Nutr Rev* 2003;61(1):1–33.

Nagata, C., et al. Soy product intake is inversely associated with serum homocysteine level in premenopausal Japanese women. *J Nutr* 2003;133(3):797–800.

Ogura, C.H., et al. Prevalence of senile dementia in Okinawa. *Intl J Epidem* 1995;24:373–80.

Sass, L. *The New Soy Cookbook*. San Francisco, CA: Chronicle Books, 1998.

Setchell, K.D.R., et al. Bioavailability, disposition, and dose-response effects of soy isoflavones when consumed by healthy women at physiologically typical dietary intake. *J Nutr* 2003; 133(4):1027–35.

Shurtleff, W., et al. *The Book of Tofu*. Berkeley, CA: Ten Speed Press, 1998, 21.

Steinberg, F.M., et al. Soy protein with isoflavones has favorable effects on endothelial function that are independent of lipid and antioxidant effects in healthy postmenopausal women. *Am J Clin Nutr* 2003;78(1):123–30.

Spinach

Beatty, S., et al. Macular pigment and age-related macular degeneration. *Br J Ophthalmol* 1999;83:857–77.

Brown, L., et al. A prospective study of carotenoids intake and risk of cataract extraction in US men. *Am J Clin Nutr* 1999;70(4):517–24.

Booth, S.L., et al. Dietary intake and adequacy of vitamin K. *J Nutr* 1998;128(5):785–8.

Castenmiller, J.J.M., et al. The food matrix of spinach is a limiting factor in determining the bioavailability of b-carotene and to a lesser extent of lutein in humans. *J Nutr* 1999;129:349–55.

Chasan-Taber, L., et al. A prospective study of carotenoids and vitamin A intakes and risk of cataract extraction in US women. *Am J Clin Nutr* 1999;70(4):431–2.

Colditz, G.A., et al. Increased green and yellow vegetable intake and lowered cancer deaths in an elderly population. *Am J Clin Nutr* 1985;41:32–6.

Ernster, L., et al. Biochemical, physiological and medical aspects of ubiquinone function. *Biochem Biophys Acta* 1995;1271: 195–204.

Greenway, H.T., et al. Fruit and vegetable micronutrients in diseases of the eye. In: *Vegetables, Fruits, and Herbs in Health Promotion.* Watson, R.R., ed. CRC Press; 2001:85–98.

Hammond, B.R., et al. Macular pigment density is reduced in obese subjects. *Invest Ophthalmol Vis Sci* 2002;43:47–50.

Handelman, G.J., et al. Lutein and zeaxanthin concentrations in plasma after dietary supplementation with egg yolk. *Am J Clin Nutr* 1999;70(2):247–51.

Hu, F.B., et al. A prospective study of egg consumption and risk of cardiovascular disease in men and women. *JAMA* 1999; 281(15):1387–94.

John, J.H., et al. Effects of fruit and vegetable consumption on plasma antioxidant concentrations and blood pressure: a randomized controlled trial. *Lancet* 2002;359(9322):1969–74.

Klein, R., et al. The association of cardiovascular disease with the long-term incidence of age-related maculopathy: the Beaver Dam Eye Study. *Ophthalmol* 2003;110(4):636–43.

Landvik, S.V., et al. Alpha-Lipoic acid in health and disease. In: *Antioxidant Status, Diet, Nutrition and Health.* Papas, A.M., ed. CRC Press, 1999:591–600.

Pratt, S. Dietary prevention of age-related macular degeneration. *J Am Optom Assoc* 1999;70(1):39–47.

Richer, S. Lutein—an opportunity for improved eye health. *JANA* 2001;4(2):6–7.

Seddon, J.M., et al. Dietary carotenoids, vitamins A, C, and E, and advanced age-related macular degeneration. Eye Disease Case-Control Study Group. *JAMA* 1994;272(18):1413–20.

Shao, A. The role of lutein in human health. *JANA* 2001; 4(2):8–24.

Simopoulos, A.P., et al. Common purslane: a source of omega-3 fatty acids and antioxidants. *J Am Coll Nutr* 1992;11(4): 374–82.

Slattery, M.L., et al. Carotenoids and colon cancer. *Am J Clin Nutr* 2000;71(2):575–82.

Steenge, G.R., et al. Betaine supplementation lowers plasma homocysteine in healthy men and women. *J Nutr* 2003; 133(5):1291–5.

Zeisel, S.H., et al. Concentrations of choline-containing compounds and betaine in common foods. *J Nutr* 2003;133(5): 1302–7.

Tea

Ahmad, N., et al. Antioxidants in chemoprevention of skin cancer. *Curr Probl Dermatol* 2001;29:128–39.

Ahmad, N., et al. Green tea polyphenols and cancer: biologic mechanisms and practical implications. *Nutr Rev* 1999;57(3): 78–83.

Arab, L. Tea and prevention of prostate, colon and rectal cancer. Third international scientific symposium on tea and human health: role of flavonoids in the diet. Washington, DC: United States Department of Agriculture, September 23, 2002.

Bell, S.J., et al. A functional food product for the management of weight. *Crit Rev Food Sci Nutr* 2002;42(2):163–78.

Benzie, I., et al. Consumption of green tea causes rapid increase in plasma antioxidant power in humans. *Nutr Cancer* 1999;34: 83–87.

Cao, Y., et al. Antiangiogenic mechanisms of diet-derived polyphenols. *J Nutr Biochem* 2002;13(7):380–90.

Chung, F. Tea in cancer prevention: studies in animals and humans. Third international scientific symposium on tea and human health: role of flavonoids in the diet. Washington, DC: U.S. Department of Agriculture, September 23, 2002.

Duffy, S.J., et al. Short- and long-term black tea consumption reverses endothelial dysfunction in patients with coronary artery disease. *Circulation* 2001;104:151–6.

Geleijnse, J.M., et al. Inverse association of tea and flavonoid intakes with incident myocardial infarction: the Rotterdam Study. *Am J Clin Nutr* 2002;75(5):880–6.

Geleijnse, J.M., et al. Tea flavonoids may protect against atherosclerosis: the Rotterdam Study. *Arch Intern Med* 1999;159(18): 2170–4.

Hakim, I. Tea and cancer: epidemiology and clinical studies. Third international scientific symposium on tea and human health: role of flavonoids in the diet. Washington, DC: United States Department of Agriculture, September 23, 2002.

Hegarty, V.M., et al. Tea drinking and bone mineral density in older women. *Am J Clin Nutr* 2000;71:1003–7.

McKay, D., et al. The role of tea in human health: an update. *J Am Coll Nutr* 2002;21(1):1–13.

Mukamal, K., et al. Tea consumption and mortality after acute myocardial infarction. *Circulation* 2002;105:2476–81.

Nakayama, M., et al. Inhibition of influenza virus infection by tea. *Lett Appl Microbiol* 1990;11:38–40.

Olthof, M.R., et al. Chlorogenic acid, quercetin-3-rutinoside and black tea phenols are extensively metabolized in humans. *J Nutr* 2003;133(6):1806–14.

Serafini, M., et al. In vivo antioxidant effect of green and black tea in man. *Eur J Clin Nutr* 1996;50:28–32.

van het Hof, K.H., et al. Bioavailability of catechins from tea: the effect of milk. *Eur J Clin Nutr* 1998;52:356–9.

Yang, C.S., et al. Effects of tea consumption on nutrition and health. *J Nutr* 2000;130(10):2409–12.

Zhu, Q.Y., et al. Regeneration of α-tocopherol in human low-density lipoprotein by green tea catechin. *J Agric Food Chem* 1999;47:2020–5.

Tomatoes

Beecher, G.R. Nutrient content of tomatoes and tomato products. *Proc Soc Exp Biol Med* 1998;218:98–100.

Edwards, A.J., et al. Consumption of watermelon juice increases plasma concentrations of lycopene and beta-carotene in humans. *J Nutr* 2003;133(4):1043–50.

Ford, E.S., et al. Serum vitamins, carotenoids, and angina pectoris: findings from the National Health and Nutrition Examination Survey III. *Annals of Epidemiol* 2000;10(2):106–16.

Fuhrman, B., et al. Lycopene synergistically inhibits LDL oxidation in combination with vitamin E, glabridin, rosmarinic acid, carnosic acid, or garlic. *Antioxid Redox Signal* 2000; 2:491–506.

Gartner, C., et al. Lycopene is more bioavailable from tomato paste than from fresh tomatoes. *Am J Clin Nutr* 1997;66: 116–22.

Giovannucci, E. Tomatoes, tomato-based products, lycopene, and cancer: review of the epidemiologic literature. *J Natl Cancer Inst* 1999;91(4):317–31.

Hadley, C.W., et al. The consumption of processed tomato products enhances plasma lycopene concentrations in association with a reduced lipoprotein sensitivity to oxidative damage. *J Nutr* 2003;133(3):727–32.

Khachik, F., et al. Lutein, lycopene, and their oxidative metabolites in chemoprevention of cancer. *J Cell Biochem* 1996; 22(suppl):236–46.

Ribaya-Mercado, J.D., et al. Skin lycopene is destroyed preferentially over beta-carotene during ultraviolet irradiation in humans. *J Nutr* 1995;125(7):1854–9.

Riso, P., et al. Tomatoes and health promotion. In: *Vegetables, Fruits, and Herbs in Health Promotion*. Watson, R.R., ed. CRC Press;2001:45–70.

Rissanen, T.H., et al. Low serum lycopene concentration is associated with an excess incidence of acute coronary events and stroke: the Kuopio Ischaemic Heart Disease Risk Factor Study. *Br J Nutr* 2001;85(6):749–54.

Snowdon, D.A., et al. Antioxidants and reduced functional capacity in the elderly: findings from the Nun Study. *J Gerontol Med Sci* 1996;51A(1):M10–6.

Stahl, W., et al. Carotenoid mixtures protect multilamellar liposomes against oxidative damage: synergistic effects of lycopene and lutein. *FEBS Lett* 1998;427:305–8.

Stahl, W., et al. Dietary tomato paste protects against ultraviolet light-induced erythema in humans. *J Nutr* 2001;131(5): 1449–51.

Stewart, A.J., et al. Occurrence of flavonols in tomatoes and tomato-based products. *J Agric Food Chem* 2000;48:2663–9.

Upritchard, J.E., et al. Effect of supplementation with tomato juice, vitamin E, and vitamin C on LDL oxidation and products of inflammatory activity in type 2 diabetes. *Diabetes Care* 2000;23(6):733–8.

Zhang, S., et al. Measurement of retinoids and carotenoids in breast adipose tissue and a comparison of concentrations in breast cancer cases and control subjects. *Am J Clin Nutr* 1997;66:626–32.

Turkey (Skinless Breast)

Anderson, J.J.B., et al. High protein meals, insular hormones and urinary calcium excretion in human subjects. In: *Osteoporosis*. Christiansen, C., Johansen, J.S., Riis, B.D., eds. Viborg, Denmark: Nørhaven A/S, 1987;240–5.

Bell, J., et al. Elderly women need dietary protein to maintain bone mass. *Nutr Rev* 2002;60(10, part I):337–41.

Bingam, S.A. High meat diets and cancer risk. *Proc Nutr Soc* 1999;58(2):243–8.

Clark, L.C., et al. Effects of selenium supplementation for cancer prevention in patients with carcinoma of the skin: a randomized controlled trial. Nutritional Prevention of Cancer Study Group. *JAMA* 1996;276:1957–63.

Cordain, L., et al. Plant-animal subsistence ratios and macronutrient energy estimations in worldwide hunter-gatherer diets. *Am J Clin Nutr* 2000;71:682–92.

Eaton, S.B., et al. An evolutionary perspective enhances understanding of human nutritional requirements. *J Nutr* 1996; 126:1732–40.

Eaton, S.B., et al. Paleolithic nutrition: a consideration of its nature and current implications. *N Engl J Med* 1985;312:283–9.

Eaton, S.B., et al. Paleolithic nutrition revisited: a twelve-year retrospective on its nature and implications. *Eur J Clin Nutr* 1997;51:207–16.

Eisenstein, J., et al. High-protein weight-loss diets: are they safe and do they work? A review of the experimental and epidemiologic data. *Nutr Rev* 2002;60(7):189–200.

Fung, T., et al. Major dietary patterns and the risk of colorectal cancer in women. *Arch Int Med* 2003;163(3):309–14.

Hamer, D.H., et al. From the farm to the kitchen table: the negative impact of antimicrobial use in animals on humans. *Nutr Rev* 2002;60(8):261–4.

Morris, M.C., et al. Dietary fats and the risk of incident Alzheimer disease. *Arch Neurol* 2003;60(2):194–200.

Norat, T., et al. Meat consumption and colorectal cancer: a review of epidemiologic evidence. *Nutr Rev* 2001;59(2):37–47.

Norrish, A.E., et al. Heterocyclicamine content of cooked meat and risk of prostate cancer. *J Natl Cancer Inst* 1999;91(23): 2038–44.

O'Dea, K. Traditional diet and food preferences of Australian aboriginal hunter-gatherers. *Philos Trans R Soc Lond B Biol Sci* 1991;334:233–41.

Pawlosky, R.J., et al. Effects of beef- and fish-based diets on the kinetics of n-3 fatty acid metabolism in human subjects. *Am J Clin Nutr* 2003;77(3)565–72.

Peregrin, T. Limiting the use of antibiotics in livestock: helping your patients understand the science behind the issue. *J Am Diet Assoc* 2002;102(6):768.

Promislow, J., et al. Protein consumption and bone mineral density in the elderly: the Rancho Bernardo study. *Am J Epidemiol* 2002;155:636–44.

Thorogood, M., et al. Risk of death from cancer and ischemic heart disease in meat and non-meat eaters. *BMJ* 1994;308: 1667–70.

Walnuts

Ahsan, S.K. Magnesium in health and disease. *J Pak Med Assoc* 1998;48:246–50.

Albert, C.M., et al. Nut consumption and decreased risk of sudden cardiac death in the Physicians' Health Study. *Arch Intern Med* 2002;162(12):1382–7.

Albert, C.M., et al. Nut consumption and the risk of sudden and total cardiac death in the Physicians' Health Study. *Circulation* 1999;98(suppl I):I–582.

Ascherio, A., et al. Intake of potassium, magnesium, calcium and fiber and risk of stroke among US men. *Circulation* 1998; 98:1198–1204.

Awad, A.B., et al. Phytosterols as anticancer dietary components: evidence and mechanism of action. *J Nutr* 2000;130:2127–30.

Devaraj, S., et al. γ-Tocopherol, the new vitamin E? *Am J Clin Nutr* 2003;77(3):530–1.

Dixon, L.B., et al. Choose a diet that is low in saturated fat and cholesterol and moderate in total fat: subtle changes to a familiar message. *J Nutr* 2001;131(2S-1):510S–26S.

Feldman, E.B. The scientific evidence for a beneficial health relationship between walnuts and coronary heart disease. *J Nutr* 2002;132(5):1062S–1101S.

Garg, M.L., et al. Macadamia nut consumption lowers plasma total and LDL cholesterol levels in hypercholesterolemic men. *J Nutr* 2003;133:1060–3.

Hu, F.B., et al. Dietary fat intake and the risk of coronary heart disease in women. *N Engl J Med* 1997;337:1491–9.

Jiang, R., et al. Nut and peanut butter consumption and risk of type 2 diabetes in women. *JAMA* 2002;288(20):2554–60.

Kris-Etherton, P.M., et al. High-monounsaturated fatty acid diets lower both plasma cholesterol and tricylglycerol concentrations. *Am J Clin Nutr* 1999;70:1009–15.

Kris-Etherton, P.M., et al. Nuts and their bioactive constituents: effects on serum lipids and other factors that affect disease risk. *Am J Clin Nutr* 1999;70(suppl):504S–11S.

Kris-Etherton, P.M., et al. Recent discoveries in inclusive food-based approaches and dietary patterns for reduction in risk for cardiovascular disease. *Curr Opin Lipidol* 2002;13(4): 397–407.

Liu, M., et al. Mixed tocopherols inhibit platelet aggregation in humans: potential mechanisms. *Am J Clin Nutr* 2003;77(3): 700–50.

Lovejoy, J.C., et al. Effect of diets enriched in almonds on insulin action and serum lipids in adults with normal glucose tolerance or type 2 diabetes. *Am J Clin Nutr* 2002;76(5):1000–6.

Morgan, W.A., et al. Pecans lower low-density lipoprotein cholesterol in people with normal lipid levels. *J Am Diet Assoc* 2000;100:312–8.

Morris, M.C., et al. Dietary intake of antioxidant nutrients and the risk of incident Alzheimer disease in a biracial community study. *JAMA* 2002;287(24):3223–9.

Ostlund, R.E., et al. Effects of trace components of dietary fat on cholesterol metabolism: phytosterols, oxysterols, and squalene. *Nutr Rev* 2002;60(11):349–59.

Sabaté, J. Nut consumption, vegetarian diets, ischemic heart disease risk, and all-cause mortality: evidence from epidemiologic studies. *Am J Clin Nutr* 1999;70(suppl):500S–3S.

Spiller, G.A., et al. Nuts and plasma lipids: an almond-based diet lowers LDL-C while preserving HDL-C. *J Am Coll Nutr* 1998;17:285–90.

Stewart, J.R., et al. Resveratrol: a candidate nutritional substance for prostate cancer prevention. *J Nutr* 2003;133(7S):2440S–3S.

Venho, B., et al. Arginine intake, blood pressure, and the incidence of acute coronary events in men: the Kuopio Ischaemic Heart Disease Risk Factor Study. *Am J Clin Nutr* 2002; 76(1):359–64.

Watkins, T.R., et al. Tocotrienols: biological and health effects. In: *Antioxidant Status, Diet, Nutrition and Health.* Papas, A.M., ed. CRC Press, 1999:479–96.

Yogurt

Bornet, F.R.J, et al. Immune-stimulating and gut health-promoting properties of short-chain fructo-oligosaccharides. *Nutr Rev* 2002;60(10, part I):326–34.

Chan, J.M., et al. Dairy products, calcium, and prostate cancer risk in the Physicians' Health Study. *Am J Clin Nutr* 2001; 74(4):549–54.

Duggan, C., et al. Protective nutrients and functional foods for the gastrointestinal tract. *Am J Clin Nutr* 2002;75(5):789–808.

Fortes, C., et al. Diet and overall survival in a cohort of very elderly people. *Epidemiol* 2000;11(4):440–5.

Gill, H.S., et al. Enhancement of immunity in the elderly by dietary supplementation with the probiotic *Bifidobacterium lactis* HN019. *Am J Clin Nutr* 2001;74(6):833–9.

Gluck, U., et al. Ingested probiotics reduce nasal colonization with pathogenic bacteria (Staphylococcus aureus, Streptococcus pneumoniae, and beta-hemolytic streptococci). *Am J Clin Nutr* 2003;77(2):517–20.

Heaney, R.P., et al. Effect of yogurt on a urinary marker of bone resorption in postmenopausal women. *J Am Diet Assoc* 2002; 102:1672–4.

Hilton, E., et al. Ingestion of yogurt containing Lactobacillus acidophilus as prophylaxis for candidal vaginitis. *Ann Intern Med* 1992;116(5):353–7.

Hooper, L.V., et al. How host-microbial interactions shape the nutrient environment of the mammalian intestine. In: *Annual Review of Nutrition*, vol. 22, 2002. McCormick, D.B., Bier, D., Cousins, R.J., eds., Palo Alto, CA: Annual Reviews;283–308.

Isolauri, E. Probiotics in human disease. *Am J Clin Nutr* 2001;73(6, suppl):1142S–6S.

Jelen, P., et al. Functional milk and dairy products. In: *Functional Foods Biochemical & Processing Aspects.* Mazza, G., ed., Lancaster, PA: Technomic Publishing Company, Inc., 1998:357–80.

Kaur, N., et al. Applications of inulin and oligofructose in health and nutrition. *J Biosci* 2002;27(7):703–14.

Kent, K.D., et al. Effect of whey protein isolate on intracellular glutathione and oxidant-induced cell death in human prostate epithelial cells. *Toxicol In Vitro* 2003;17(1):27–33.

Liska, D., et al. Gut restoration and chronic disease. *JANA* 2002; 5(4):20–33.

Madden, J.A., et al. A review of the role of the gut microflora in irritable bowel syndrome and the effects of probiotics. *Brit J Nutr* 2002;88(suppl 1):S67–S72.

Mann, G.V., et al. Studies of a surfactant and cholesteremia in the Maasai. *J Nutr* 1974;27:464–9.

Mitral, B.K., et al. Anticarcinogenic, hypo-cholesterolemic, and antagonistic activities of lactobacillus acidophilus. *Crit Rev Microbiol* 1995;21(3):175–214.

Perdigon, G., et al. Antitumour activity of yogurt: study of possible immune mechanisms. *J Dairy Res* 1998;65(1):129–38.

Rachid, M.M., et al. Effect of yogurt on the inhibition of an intestinal carcinoma by increasing cellular apoptosis. *Int J Immunopathol Pharmacol* 2002;15(3):209–16.

Saavedra, J.M. Clinical applications of probiotic agents. *Am J Clin Nutr* 2001;73(6,suppl):1147S–51S.

Salminen, S., et al. Demonstration of safety of probiotics—a review. *Int J Food Microbiol* 1998;44:93–106.

Sanders, M.E. Probiotics: considerations for human health. *Nutr Rev* 2003;61(3):91–9.

Seppo, L., et al. A fermented milk high in bioactive peptides has a blood pressure-lowering effect in hypertensive subjects. *Am J Clin Nutr* 2003;77(2):326–30.

Teitelbaum, J.E, et al. Nutritional impact of pre- and probiotics as protective gastrointestinal organisms. In: *Annual Review of Nutrition*, vol. 22, 2002. McCormick, D.B., Bier, D., Cousins, R.J., eds., Palo Alto, CA: Annual Reviews, 107–38.

Wong, C.W., et al. Immunomodulatory effects of dietary whey proteins in mice. *J Dairy Res* 1995;62(2):359–68.

Index

beans (continued)
  canned, 46–47, 53, 330
  coronary heart disease and,
      51–54
  fiber and, 53–54
  health benefits of, 46, 48–51
  history of, 47–48
  low-fat protein in, 48–51
  nutrients in, 45
  obesity and, 55
  reducing gas caused by, 47
  in roasted red pepper
      hummus, 286–87
  varieties of, 56–57
  white, soup, with greens and
      rosemary, 280–81
beef brown sauce, 249–50
bell peppers, orange, 171
berry(ies):
  crisp with nuts and oatmeal
      topping, 257–58
  packaged fresh, shopping list
      for, 345–46
  see also blueberry(ies)
beta-carotene, 20, 27, 29, 192,
    194, 198
  in pumpkin, 119–24
  recommended daily intake of,
      319
  in spinach, 173
beta cryptoxanthin, 320
beta glucan, 87
betaine, 167
biotin, 217
birth defects, 75, 80, 225
black beans, 56
black tea, 182–85
blood coagulation, 173
blood sugar, 54, 87, 106, 110, 175
blueberry(ies), 59–73
  antioxidants in, 60–63, 66–67,
      70

bread, Golden Door, 265–66
French Paradox and, 67–68
health benefits of, 59–67
nutrients in, 59–61
polyphenols in, 61–64
pops, frozen yogurt, 73
ricotta torte with, 292–93
and yogurt shake, 240
bone health, 75, 80–81, 155, 188
bread, 328
  Golden Door blueberry,
      265–66
  whole grain toast with almond
      butter, 260
breakfast, 90, 239
  broccoli frittata, 275–77
  Rancho La Puerta crisp,
      283–85
  recipe, day one, 245
  recipe, day two, 252
  sweet potato pie, 267–69
  sweet potato scones, 293–94
  tropical fruit yogurt parfait,
      259–60
  whole grain toast with almond
      butter, 260
broccoli, 74–84
  buying and cooking of, 81–84
  cancer and, 76–79
  cardiovascular health and, 75,
      81
  frittata, breakfast, 275–77
  health benefits of, 75–76,
      79–81
  nutrients in, 74–75
  raw vs. cooked, 76
  sprouts, 79
  sulfur compounds in, 78
brown rice, 98
buckwheat, 98
  honey glaze, roasted butternut
      squash with, 129

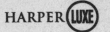